Designing a structured cabling system to
ISO 11801 2nd edition

# Designing a
## Structured Cabling
## System to
## ISO 11801 2nd Edition

Cross-Referenced to European CENELEC
and American Standards

### Barry J. Elliott

*Brand-Rex Ltd.*
*Glenrothes, Fife, Scotland*

MARCEL DEKKER, INC.          NEW YORK · BASEL

ISBN: 0-8247-4130-7

Headquarters
Marcel Dekker, Inc.
270 Madison Avenue, New York, NY 10016
tel: 212-696-9000; fax: 212-685-4540
http://www.dekker.com

Marcel Dekker, Inc., offers discounts on this book when ordered in bulk quantities. For more information, write to Special Sales/Professional Marketing at the headquarters address above.

First published 2002

Woodhead Publishing Limited
Abington Hall, Abington, Cambridge CB1 6AH, England
www.woodhead-publishing.com

Typeset by SNP Best-set Typesetter, Ltd., Hong Kong
Printed by TJ International, Padstow, Cornwall, England

PRINTED IN THE UNITED KINGDOM

# Contents

# Preface

This book is presented as an aid for information technology (IT) managers, consultants, cable installation engineers and system designers who need to understand the technology of the subject and the vast panoply of standards that accompany it. The book is a design manual for structured cabling and explains the terminology and physics behind the standards, what the relevant standards are, how they fit together and where to obtain them. Anybody studying this book will be able to read the standards, understand manufacturers' data sheets and their conflicting claims and be suitably equipped to address those problems raised by the need to design, procure, install and test correctly a modern cabling system, using both copper and optical fibre cable technology.

The book is based upon the design recommendations of ISO/IEC 11801: *Information Technology – generic cabling for customer premises,* 2nd edition 2002. But this is only part of the story: ISO 11801 also references over a hundred other standards concerning product specification, EMC/EMI, testing, administration and cable containment. In many cases the standards give a range or set of design options that the reader can take, but with little advice on which route to pick. This book will endeavour to guide the reader around the standards and lead to a cable system design methodology suitable for all premises and campus cabling installations.

For those interested in some of the deeper aspects of the subject then the author recommends *Cable Engineering for Local Area*

*Networks*, by Barry J Elliott, Woodhead Publishing Ltd, Cambridge, 2000, www.woodhead-publishing.com, ISBN 1 85573 488 5 (ISBN 0 8247 0525 4 in North America).

*Barry Elliott*
*Credo ut intelligam*

*For Siobhan, Sean and Francesca*

# List of abbreviations

| | |
|---|---|
| ACR | Attenuation to Crosstalk Ratio |
| ADC | Analogue to Digital Converter |
| ADSL | Asymmetric Digital Subscriber Line |
| ANSI | American National Standards Institute |
| ASTM | American Society for Testing and Materials |
| ATM | Asynchronous Transfer Mode |
| AWG | American Wire Gauge |
| BER | Bit Error Rate |
| BERT | Bit Error Rate Tester |
| BN | Bonding Network |
| CAD | Computer Aided Design |
| CAM | Computer Aided Manufacturing |
| CATV | Community Antenna Television (Cable TV) |
| CCTV | Closed Circuit Television |
| CBN | Common Bonding Network |
| COA | Centralised Optical Architecture |
| CP | Consolidation Point |
| CSA | Canadian Standards Association |
| dB | decibel |
| EIA | Electronic Industries Alliance |
| ELFEXT | Equal Level Far End Crosstalk |
| EMC | Electro Magnetic Compatibility |
| EMI | Electro Magnetic Immunity (or sometimes 'EM *Interference*') |

| EN | European Norm |
|----|---------------|
| ESD | Electro Static Discharge |
| ETSI | European Telecommunications Standards Institute |
| EU | European Union |
| FCC | Federal Communications Commission |
| FD | Floor Distributor |
| FDDI | Fibre Distributed Data Interface |
| FEXT | Far End Crosstalk |
| FOCIS | Fiber Optic Connector Intermateability Standards |
| FTP | Foil screened Twisted Pair |
| FTTD | Fibre-to-the-Desk |
| GHz | giga Hertz |
| HVAC | Heating, Ventilation and Air Conditioning |
| IBS | Intelligent Building Systems |
| ICEA | Insulated Cable Engineers Association |
| IDC | Insulation Displacement Connector |
| IEC | International Electro Technical Commission |
| IEE | Institute of Electrical Engineers (UK) |
| IEEE | Institute of Electrical and Electronic Engineers (USA) |
| IP | Internet Protocol |
| ISDN | Integrated Services Digital Network |
| ISO | International Organisation for Standardisation |
| ITU | International Telecommunications Union |
| LAN | Local Area Network |
| LED | Light Emitting Diode |
| MAN | Metropolitan Area Network |
| Mb/s | Megabits per second |
| MHz | megahertz |
| MUTO | Multi User Telecommunications Outlet |
| NEC | National Electrical Code (USA) |
| NEMA | National Electrical Manufacturers Association (USA) |
| NEXT | Near End Crosstalk |
| NFPA | National Fire Protection Association (USA) |
| NIC | Network Interface Card |
| nm | nanometres |
| PABX | Private Automatic Branch Exchange |
| PAM | Pulse Amplitude Modulation |

| PAN | Personal Area Network |
| PBX | Private Branch Exchange |
| POF | Plastic Optical Fibre |
| PS-ACR | Power Sum Attenuation to Crosstalk Ratio |
| PS-ELFEXT | Power Sum Equal Level Far End Crosstalk |
| PS-NEXT | Power Sum Near End Crosstalk |
| PSTN | Public Switched Telephone Network |
| SAN | Storage Area Network |
| ScTP | Screened Twisted Pair |
| SFF | Small Form Factor (optical connectors) |
| S-FTP | Screened, Foil Twisted Pair |
| SONET | Synchronous Optical Network |
| STP | Screened (or shielded) Twisted Pair |
| TGB | Telecommunications Ground Busbar |
| TGMB | Telecommunications Ground Main Busbar |
| TIA | Telecommunications Industry Association |
| TO | Telecommunications Outlet |
| TOC | Terminated Open Circuit |
| TP | Transition Point |
| TSB | Telecommunications Systems Bulletin |
| UL | Underwriters Laboratories Inc (USA) |
| UTP | Unscreened (or unshielded) Twisted Pair |
| VCSEL | Vertical Cavity Surface Emitting Laser |
| WAN | Wide Area Network |
| WDM | Wavelength Division Multiplexing |
| xDSL | Digital Subscriber Link ('x' can denote different methods) |

# 1

# Introduction

Structured cabling, sometimes called premises cabling, arose as a new subject in the early 1980s after it became apparent that the growth of the information technology (IT) business could be slowed by the wide range of incompatible cables and network interfaces that had grown up with the computer mainframe industry of the 1960s and 1970s. When computers were first generally deployed it was the practice of every manufacturer to design their own cables specifically for a certain range of computers. There were a few standards such as RS 232 but more often companies like IBM would specify a cable for every computer system in their product range. Cables and connectors could be any combination of twisted pairs, multicores, coaxial cables or even the 'new fangled' optical fibres.

The network architecture could be tree, bus, star, point-to-point or daisy-chain. Point-to-point means that a cable is deployed from the central processor directly to, and dedicated solely to, the peripheral.

Daisy-chain means a cable running from the central processor to the first peripheral and then linking on to the next peripheral and so on. Some form of addressing or polling protocol is required to allow each peripheral to know when it is being addressed and for the central processor to be able to recognise which peripheral has communicated with it.

The disadvantages of this dedicated form of wiring are easy to see. The problems include a dedicated and proprietary cable for every manufacturer and moves and changes are costly and time consuming as provision for future expansion is rarely implemented.

A building of the 1970s or 1980s may have had many different types of cables for different applications, such as numerous mainframe computers from IBM, ICL, Honeywell, Sperry and so on. There would also be two-, three-, four- or even five-pair cables for telephones and PBX extensions. There might also be 25-, 50-, 100- and 300-pair cables for telephony backbone cabling; 75-$\Omega$ coax for CCTV/security video; electronic point of sale (EPOS) cabling and various cables for security, access control, heating and ventilation control (HVAC), public address system, environmental monitoring, smoke and fire detection and fire alarms.

Local area networks (LANs), a standardised way of interconnecting computers from different manufacturers, could not have evolved without a common set of cables and cabling to transport that data in a reliable and predictable way. Today, all data applications run on the structured cabling system as do most, but not all, PBX telephone systems. All video applications can run on the structured cabling system if it is of a high enough 'Category' and bandwidth.

The building control systems, generally grouped under the heading IBS or intelligent building systems, have recently started to utilise the benefits of structured cabling but at a rate slower than most anticipated. Some fire detection, alarm and emergency lighting circuits are and will be required to use separate fire survival cable such as mineral insulated cable.

IBM is generally credited with the introduction of structured cabling around 1984. IBM proposed that a set of data cables should follow the same philosophy as telephone cabling, that is, a common set of flood-wired cables that terminate at every position where a user sits or even where somebody may sit in the future. All the cables would terminate in a common connector at the user end and at a set of patch panels at the administration end. By means of the patch panel any user could be connected to any application. The initial investment may be more but the lower cost of moves and changes would give the structured cabling system a payback period of between three and five years.

The IBM Cabling System (sometimes referred to as ICS) utilised a twin pair screened cable of 150-$\Omega$ characteristic impedance. A brand new hermaphroditic connector was designed to complement the cable, and the application firmly in the designers' minds at the time was the IBM Token Ring LAN.

Token Ring was launched as a 4 Mb/s LAN although a 16 Mb/s version soon followed. The IBM Cabling System however has a frequency performance of at least 300 MHz, and together with the screening elements and large connector it soon became apparent that it was relatively over-engineered for the job it had to do. The IBM Cabling System has since matured into the Advanced Connectivity System (ACS) that offers standards-based, 100-$\Omega$ connectivity.

By 1988 a part of AT&T, later trading as Lucent Technologies and then Avaya, advanced the idea of using simple four-pair, 100-$\Omega$, unshielded twisted pair (UTP) cabling direct from the American telephone network. One-ten cross-connects and the 8-pin connector we now refer to as the RJ45 were pressed into service and used to demonstrate at least 1 Mb/s operation but at a much lower cost than the IBM Cabling System. Like ICS, AT&T's premises distribution system could link up 'legacy' computer and cabling systems by the use of a balun, a device that converts from one cable type and impedance to another. The AT&T system formed the basis of the 100-$\Omega$, four-pair cabling system still in use today for generic premises cabling systems.

By 1990, many other cable and connector manufacturers were offering look-alike cabling systems all claiming widely differing performance. To offer some guidance to customers and suppliers, the American distribution company Anixter introduced a grading system for the cable known as 'Levels' with Level 3 being the highest.

Ethernet came to the world's attention in 1976, described in a paper by Boggs and Metcalfe. The cable used is a large bright yellow coaxial cable used in a system that came to be known as 10BASE-5. The '10' meaning the speed in Mb/s; 'BASE' meaning baseband operation, that is, the signal goes from zero Hertz upwards without being modulated onto a higher frequency carrier. The '5' meant that the cable could be up to 500 m long, although many believed it referred to the 50-$\Omega$ characteristic impedance of the cable. The cable had to be terminated with 50-$\Omega$ resistors at either end to stop signals reflecting up and down the cable and users could plug into the cable at any point with a special 'vampire-tap' connector. This was followed shortly after by a smaller coaxial cable based version called 10BASE-2, or Cheapernet or Thin-net. Neither version followed the ideal of structured cabling whereby all users are star-wired back to some central point.

Ethernet, and thereby most of the LAN market, started to converge with the nascent structured cabling market in 1992 with the introduction of the 10BASE-T Ethernet standard.

Although the first forms of Ethernet used a high quality coax, 10BASE-T was designed to work on good quality but basic telephone cabling, hence the 'T'. 10BASE-T and structured cabling went hand-in-hand and the market accepted the whole concept and benefits of structured cabling within three years.

By the early 1990s the various bodies that write standards started to catch up with events, specifically ANSI, who gave the job to the TIA (USA) and EIA. The TIA and EIA changed the word 'Level' to 'Category' and Category 3 was born. The TIA, in association with the EIA and ANSI published TSB 36 for cable and TSB 40 for connecting hardware.

Category 3 described a cabling system with a 16-MHz bandwidth and with 10BASE-T firmly in mind. When IBM introduced their 16 Mb/s Token Ring LAN they pointed out that the cabling system needed a bandwidth of at least 20 MHz to support it. Category 4 was hurriedly rushed out offering a 20 MHz bandwidth, but it had a life measured in months as the designers already had their eye on 100 Mb/s Ethernet. Nobody quite knew what kind of coding would be required for a 100 Mb/s LAN but basic physics said that it could always be encoded within a 100 MHz bandwidth channel. Hence 100 MHz Category 5 was born and published as a Standard, TIA/EIA 568A, in 1995, closely followed by the International ISO 11801 and European CENELEC EN 50173. The cabling systems described in 1995 lasted for the rest of the decade but for the new millennium a range of advancements in structured cabling and LAN technology, such as Category 6, Category 7 and Gigabit Ethernet, are now firmly with us.

The international Standard for structured cabling is now ISO

11801, published jointly by the ISO and the IEC based in Geneva. ISO 11801 is a truly international document with members coming from all the continents except Antarctica. In the EU the relevant Standard is published by CENELEC as EN 50173, although the differences between the ISO and CENELEC standards are now very hard to spot. In many instances it is the same authors sitting on both the ISO/IEC and CENELEC committees. The Standard for the United States is TIA/EIA 568B. The equivalent Standard for Canada is CAN/CSA T529 (the Canadian Standards Association) and in Australia and New Zealand it is AS/NZS 3080.

This book addresses the design requirements for ISO 11801 but cross-references all the major items to their equivalent standards within the European CENELEC system and the American TIA/EIA series. It is important to recognise that ISO 11801 is not a stand-alone Standard. It in turn references no less than 128 other standards from a bewildering array of related technological issues. It would be impossible to buy, let alone read or study, every single referenced standard and the resulting expanding tributaries of further referenced standards. In this book all the major and relevant standards and related design issues will be covered to enable the reader to specify, install and test a modern reliable structured cabling system that will satisfy the next generation of standards, led by ISO 11801 2$^{nd}$ edition 2002.

# 2

# First pick your standards

## 2.1 Standards: what standards?

Step 1 in designing or procuring a structured cabling system is to decide which standards philosophy you want to adhere to. The choices are:

- None at all, just specify a grade of cabling and let the installer sort it all out.
- Pick a famous brand and hope that the ministrations of a large corporation will ensure that something relevant, useful and reliable is installed.
- Pick and mix from different standards.
- Use local/national standards.
- Use international standards.

Leaving the choice of components and the resulting performance totally in the hands of an installer is a risky business and should only be contemplated by those looking for a very undemanding network at the absolute lowest price. However, do not forget the ongoing cost of ownership when things start to go wrong. Many surveys point out that more than half of all network faults are cable related.

The days of picking and specifying a major brand name as the sole supplier are probably over with the advent of comprehensive international standards, although it is through the efforts of some of the major suppliers that serious standards and well-engineered products do exist. Many of the larger users now tend to 'prequalify' a group of major manufacturers' products, after specifying them according to the standards, and then prequalify a group of competent installers, telling them they can bid any product from the pre-approved list.

The pick and mix approach is not necessary either as full families of standards now exist across all the principal standards-writing bodies. A strange mix of standards usually says more about the specification compiler's lack of knowledge rather than anything else. However, nobody has a standard for everything and an open mind is needed to get the best overall package put together to address all network problems. Remember nowadays that the cabling installation does not just require an overall design standard but needs information relating to EMC/EMI, fire performance, cable containment, earthing and bonding and a host of other related subjects.

There are few local standards relating to cabling, apart from some fire regulations, but there are national standards for the USA, for Canada, for Australia and New Zealand and also for the EU.

Systems located in the USA should design cable standards to the ANSI/TIA/EIA range of standards with their supporting National Electrical Codes. The principal design standard in the USA is TIA/EIA 568B. The equivalent standard for Canada is CAN/CSA T529 and in Australia and New Zealand it is AS/NZS 3080.

In the EU the relevant standards are written by CENELEC and the standard for structured cabling is EN 50173. All CENELEC standards start with EN for European Norm.

World standards are written by ISO and IEC. Both are based in Geneva, Switzerland, but they are truly international bodies with representatives from all continents forming their committee membership. Their standards are written for a worldwide audience and can be invoked in the EU and the USA with full impunity. In the EU it is expected that CENELEC standards will be utilised where they exist, when users are spending public money. The EU Procurement Directive looks towards European standards to ensure fair and open competition across all the member states.

Many multinational organisations, regardless of location, choose to

| Table 2.1  Cabling design standards for different regions | | | | |
|---|---|---|---|---|
| European Union | United States | Canada | Australia and New Zealand | Rest of the World |
| EN 50173 | TIA/EIA 568 | CAN/CSA-T529 | AS/NZS 3080 | ISO 11801 |

use ISO/IEC standards to ensure a worldwide compatibility within their installed cabling.

ISO writes system standards, like ISO 9000, the quality Standard and of course ISO 11801, the cabling system Standard. The IEC concentrates on components and methodologies. For example, ISO 11801 invokes IEC 61156 for the cable detail and IEC 61935 for the test method detail. Both CENELEC and ANSI/TIA/EIA standards also refer to IEC standards, such as IEC 60603, the 8-pin connector standard.

This book addresses the design requirements for an ISO 11801 compliant cabling system but cross-references all the major items to their equivalent standards within the European CENELEC system and the American TIA/EIA series.

In this book we will design a cabling system according to ISO 11801 for both copper and optical cabling and focus on the areas where decisions and judgements need to be made where the standard is not totally prescriptive, for example, in the area of screened versus unscreened cabling. The book will take a wider view than just ISO 11801 as this Standard is not a complete recipe for a cabling system. We must also address testing issues, fire performance, earthing, screening and bonding, cable containment methods, cable plant administration and several other vital areas. Table 2.1 summarises the principal cabling design Standards for different regions.

## 2.2   Who writes the standards?

Every European country still maintains its own national standards body, such as the British Standards Institute in the UK, but CENELEC standards are adopted as national standards where they exist, for example the installation standard EN 50174 is published in the UK by the British Standards Institute as BS EN 50174.

Other principal standards writing bodies are described in the rest of this chapter.

**ANSI** the American National Standards Institute, publishes many Standards including many related to test methods and LANs. The TIA and EIA publish their standards under the auspices of ANSI so that they are ANSI/TIA/EIA standards. ANSI describes itself as the administrator and coordinator of the United States private sector voluntary standardisation system since 1918. ANSI does not in itself develop American National Standards (ANS) but rather facilitates development by establishing consensus among qualified groups. ANSI accredits more than 175 entities actually to develop the Standards, for example bodies such as the TIA.

**ASTM** the American Society for Testing and Materials, generates many basic electrical tests that are referenced in other more specific tests for structured cabling.

**ATM Forum** is an international non-profit organisation composed of mainly network equipment manufacturers with the aim of promoting the use of ATM (asynchronous transfer mode) networking technology by agreeing standards and common interface requirements. The ATM forum consists of a worldwide technical committee, three marketing committees and a user committee, through which ATM end users can participate. The ATM Forum was created in 1991 and now has over 600 members. Dozens of technical specifications have been published by the ATM Forum covering issues such as control signalling, LAN emulation, network management and physical layer. It is the latter that stipulates electrical and optical performance requirements of the cabling system.

**BiCSi** is a not-for-profit trade organisation dedicated to promoting professional qualifications within the cabling industry. BiCSi now operates in 70 countries with over 15000 members and offers the professional RCDD, Registered Cable Distribution Designer, qualification. BiCSi publishes many documents relating to the design, installation and testing of structured cabling systems.

**CEN** is another European body that works in partnership with CENELEC and ETSI. CEN's mission is, 'to promote voluntary technical harmonisation in Europe in conjunction with world-wide bodies and its partners in Europe.'

**EIA** the (American) Electrical Industry Alliance, writes more specific component-based electrical standards such as EIA-310, racks and cabinets. The EIA describes itself as providing a forum for industry to develop standards and publications in major technical areas: electronic components, consumer electronics, electronic information and telecommunications.

**IEE** the (British) Institute of Electrical Engineers writes the British *Wiring regulations* (also known as BS 7671), which contains safety issues concerning power cable installation.

**IEEE** the Institute of Electrical and Electronic Engineers, is a professional body whose main claim to fame in this arena is writing most of the LAN standards in use today. For example the ubiquitous Ethernet standard comes from the IEEE 802.3 committee. The IEEE is concerned with cable system performance, as it is the physical layer of the stack of network protocols. The IEEE, with its 330 000 members in 150 countries, produces more than 30% of the world's published literature in electrical engineering, computers and control technology. The IEEE also publishes the National Electrical Safety Code, NESC, which covers basic provisions for the safeguarding of persons from the hazards arising from the installation, operation or maintenance of outside plant.

**ETSI** is the European Telecommunications Standards Institute based in France. It produces voluntary telecommunications standards in response to requests from its members, currently numbering 700 across 50 countries.

**FCC** the (American) Federal Communications Commission lays down many rules regarding the use of telecommunications equipment within the United States. Two sets of rules particularly apply to the use of structured cabling systems; FCC Part 15, *Electromagnetic Radiation Issues* and FCC Part 68, *Connection of Premises Equipment and Wiring to the Telecommunications Network.*

**ITU** the International Telecommunications Union was formerly known as CCITT, and like many other standards bodies is located in Geneva, Switzerland. The ITU provides a forum within which governments and the private sector can coordinate global telecom networks and services. The ITU-T fulfils the purposes of the ITU, relating telecommunications standardisation by studying technical, operating and tariff questions and adopting 'Recommendations' (note: not called 'standards') with a view to standardising telecommunications on a worldwide basis. An example of recent ITU-T work is the UIFN or universal international freephone number.

**NEMA** the (American) National Electrical Manufacturers Association, produces cable specifications from its *High Performance Wire and Cable Section, Premises Wiring* section. The TIA adopts some of the NEMA work for inclusion into its wider system-based standard.

**NFPA** the (American) National Fire Protection Association, the NFPA, has spent the last one hundred years developing codes and standards concerning all areas of fire safety. It now has 65 000 members in 70 countries. There are currently more than 300 NFPA fire codes in use, such as:

- NFPA 1, Fire Prevention Code.
- NFPA 54, National Fuel Gas Code.
- NFPA 70, National Electrical Code.
- NFPA 101, Life Safety Code.

NFPA 70, with its National Electrical Code and various articles is the most prevalent code within the American structured cabling industry.

**TIA** the (American) Telecommunications Industry Association is a trade association active since 1924. The TIA attempts to represent the American telecommunications industry in conjunction with the Electronic Industries Alliance, the EIA. The TIA has five product-oriented divisions, User Premises Equipment, Network Equipment, Wireless Communications, Fiber Optics and Satellite Communications. Each division prepares standards dealing with performance testing and compatibility. The TIA maintains standards formulating groups. The most important one for structured cabling is TR 42, *User Premises Telecommunications Cabling Infrastructure*. Another relevant group is TR41, *User Premises Telecom Requirements*. The TIA also maintains two fibre optic divisions, FO-2 and FO-6. FO-2 group develops physical-layer Optical Fiber System Test Procedures (OFSTP) TIA-526 series. The FO-6 group develops Standard Fiber Optic Test Procedures (FOTP) TIA-455 series, Informative Test methods and Fiber Optic Connector Intermateability Standards, FOCIS.

**UL** the (American) Underwriters Laboratories Inc, or UL, is an independent, not-for-profit product safety testing and certification organisation. 'UL' marks appear on about 15 billion items every year and 90 000 new products are evaluated every year. UL develops tests such as UL 910, the plenum fire-rating test. Note that tested products can be described as 'Listed', 'Classified' or 'Recognised'.

- A product can be 'Listed' after it has successfully passed a series of electrical and mechanical tests that simulate all likely hazards to be encountered by that product.
- A product can be 'Recognised' after it has been tested and passes for use as a component in a 'Listed' package.
- A product is 'Classified' if it is evaluated and passes tests for specific hazards or certain regulatory codes.

# 3

# Topology or architecture of structured cabling systems

## 3.1 Introduction

Structured cabling systems have to conform to a very particular model that describes the topology or architecture of the cabling system. If the cabling does not conform to the ISO 11801 model then it is no longer structured cabling but some form of applications-dependent cabling. The great advantage of following the ISO 11801 design rules is that the performance of LANs, telecommunications and datacommunications systems, video and a host of other recognised communication protocols will work predictably over the cabling system and usually manufacturers will be happy to guarantee that performance if the model is followed strictly.

It is important to separate the logical connection of a LAN from the physical connection seen in the cabling system. Local area networks are sometimes designed to work in a 'bus' fashion, for example the first version of Ethernet where all users plugged into a length of coaxial cable and used it as a common bus. Some LANs worked as a loop or ring, for example Token Ring and FDDI, and they still work that way even when connected via conventional structured cabling in a 'star-wired' manner. Figure 3.1 shows the original Ethernet bus structure and Fig. 3.2 shows a ring structure.

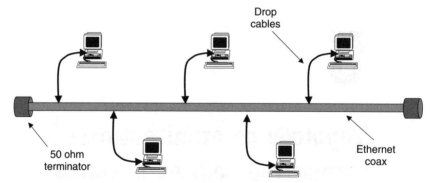

Fig. 3.1 Ethernet bus.

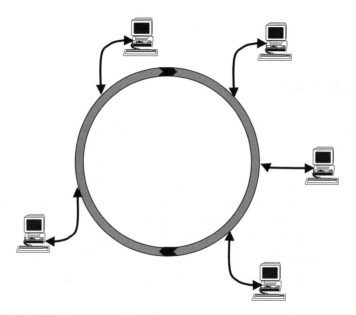

Fig. 3.2 Ring connection.

Early network architectures were either point-to-point or daisy-chain. Point-to-point means that a cable is deployed from the central processor directly to, and dedicated solely to, the peripheral, see Fig. 3.3.

'Daisy-chain' means a cable running from the central processor to the first peripheral and then linking on to the next peripheral and so

**Fig. 3.3** Point-to-point connection.

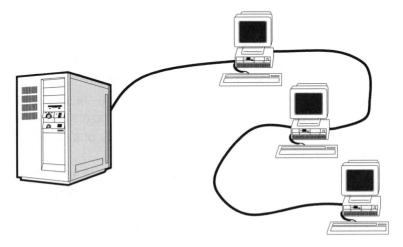

**Fig. 3.4** Daisy-chain connection.

on. Some form of addressing or polling protocol is required to allow each peripheral to know when it is being addressed and for the central processor to be able to recognise which peripheral has communicated with it, see Fig. 3.4.

The disadvantages of any dedicated forms of wiring are:

- It is dedicated and proprietary to a particular manufacturer and/or computer system.
- Moves and changes are costly and time consuming as provision for future expansion is rarely implemented.

The majority of computer systems and LAN devices today are made with the expectation that they will be connected to conventional structured cabling systems and will have a predictable level of performance if the cabling design rules have been followed.

## 3.2   Standardised structured cabling

The cabling models developed for ISO 11801 are almost identical to EN 50173 and TIA/EIA-568-B; any differences will be highlighted.

We now need to start using the terminology of structured cabling. We have cable, which can appear as the horizontal, the building backbone or the campus backbone cabling. Patch panels are known as distributors. The link between the horizontal cabling and building backbone is the floor distributor. The link between the building and campus backbone is the building distributor and where all the campus cabling comes together we have the campus distributor. We can have as many floor and building distributors as we like but we can only have one campus distributor in any one discrete structured cabling entity.

In American (US) terminology the distributors are known as cross-connects: hence we have the horizontal cross-connect, the intermediate cross-connect and the main cross-connect.

The wall or floor outlet becomes the telecommunications outlet (TO). Flexible cables, cords or patchcords from the TO linking into the active equipment are the work area cables.

ISO 11801 structured cabling conforms to the three-layered hierarchical model. This is composed of the horizontal cabling, the building backbone cabling and the campus backbone cabling. We can infer from this that there cannot be more than three layers of patching, that is, between the horizontal and building backbone cabling, between the building and campus backbone cables and where all the campus cabling finally meets. If there are more layers of patching than this then it is no longer ISO 11801 structured cabling. Figure 3.5 details the basic three-layer hierarchy.

The vast majority of installed structured cabling is made up of the horizontal cabling. This is the cable from the wall or floor outlet to the patch panel. It does not matter which direction the cable goes, it is always defined as the horizontal cabling.

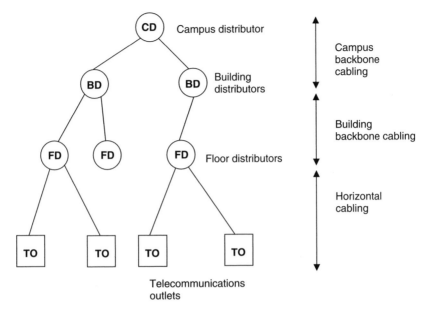

**Fig. 3.5** Three-layer hierarchical model.

In a smaller installation there might be no more than horizontal cabling; there is no requirement to use all three levels of patching available. Single building sites will probably only have the horizontal cabling and the building backbone cabling. Multibuilding sites, or campuses, will require all three levels.

The three layers can be effected in either copper or optical cables and cable selection is addressed later in this book. There is an alternative arrangement known as COA, which has also been referred to as fibre-to-the-desk (FTTD), optical home-run and collapsed backbone. These are all obviously optical fibre solutions and the benefits of optical fibre versus copper will also be considered later under cable selection. The point is that whereas high speed LANs are limited to 100-m transmission distance over copper cables, they can go much further over optical fibre. If using optical fibre, therefore, there is no real need to go through the various layers of patching seen at the horizontal/building backbone transition or even the building/campus transition. The optical fibre should be able to run all the way from the user's workstation directly back to the main equipment room. Thus

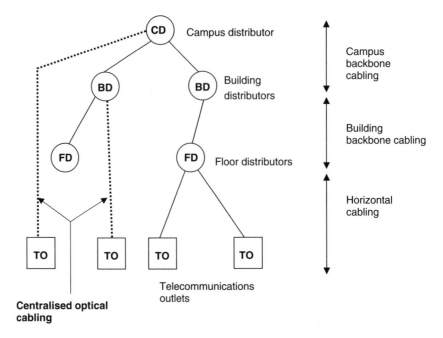

**Fig. 3.6** Centralised optical architecture, COA.

COA should be able to offer substantial savings by eliminating the horizontal/backbone cross-connects and their associated cabinets, active equipment, power supplies, floor space etc. Figure 3.6 shows the principal three-layer model with the possible bypass of the intervening patching elements by use of centralised optical fibre.

ISO 11801 and EN 50173 allow a direct run of optical fibre of up to 2000 m from the outlet to the equipment room. The American TIA/EIA-568-B standard only allows for 300 m.

Excluding COA, the maximum length of the horizontal cabling cannot exceed 90 m; this is known as the permanent link, PL. The addition of patch cords on either end of the PL forms the channel. The maximum channel distance in the horizontal is 100 m. Possible strategies to overcome this distance limitation are discussed in the design section.

The total distance made up from the addition of the horizontal plus the building backbone plus the campus backbone cabling must not exceed 2000 m. TIA/EIA-568-B is slightly different. The distance from the horizontal cross-connect to the main cross-connect cannot exceed the following distances:

*   800 m when using twisted pair cabling.
*   2000 m when using multimode fibre.
*   3000 m when using single mode fibre.

We can see therefore that up to 3.1 km would theoretically be possible under TIA/EIA 568-B rules.

The overriding rule however is to ensure that whatever communications protocol is envisaged for the future, it will work reliably over the length and type of cabling proposed. We will consider this design aspect further in Chapter 5 and at various other points in the book.

We have looked at the basic topology of the three-layer hierarchy with possible COA bypass. There are three other architectural rules we need to understand as well:

*   Direct links between distributors.
*   Consolidation points and multiuser TOs.
*   Interconnect and cross-connect model.

For reasons of security and redundancy the distributors may be directly connected to each other. Figure 3.7 illustrates this.

Depending upon distance, the type of cable used for interdistributor links would normally be the same as the adjoining backbone cables.

A consolidation point, CP, is another point of administration located at some point within the horizontal cabling and allows some flexibility between the main fixed cabling and the final links to the TOs. It may prove useful in an open office environment or where there is a large floor area served by cabling laid under false floors. TIA/EIA-568-B offers a useful definition of the CP as being a 'location for interconnection between horizontal cables extending from building pathways and horizontal cables extending into furniture pathways'.

The CP is not to be confused with the transition point, TP, which appeared in earlier standards. The original concept of the TP was to act as a joint between mechanically dissimilar cables such as flat and round cables. These options no longer exist and the TP, which was

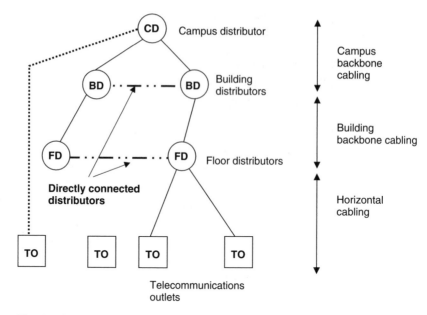

**Fig. 3.7** Directly connected distributors.

defined as *not* being a 'point of administration', no longer appears in any of the standards apart from a brief mention in TIA/EIA-568-B.

The CP may be composed of IDC (insulation displacement connector) punchdown blocks, 8-pin plug and sockets or patch panels. Some of the basic rules for using CPs are:

- The CP shall only contain passive components and shall not be used in a cross-connect format.
- A CP shall be considered as part of the administration system.
- A CP shall be placed in an accessible location.
- A CP shall serve a maximum of twelve work areas.
- When using a CP, it should be located at least 15m away from the floor distributor, when using twisted pair cable (this is to reduce the chance of return loss problems).

Figure 3.8 represents the disposition of a CP.

A similar component is the MUTO assembly. A single-user TO will have two connections, the first should be a 4-pair balanced copper

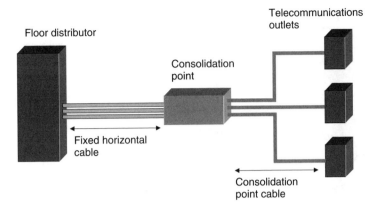

Telecommunications
outlets

Floor distributor

Consolidation
point

Fixed horizontal
cable

Consolidation
point cable

**Fig. 3.8** Consolidation points.

cable and the second should be another 4-pair cable or a pair of optical fibres.

A MUTO can serve up to 12 work areas and thus looks more like a wall mounted patch panel. Work area cabling, that is the patch cords, can now be made much longer, up to 20m, with a corresponding reduction in the length of the fixed horizontal cabling to account for the extra attenuation caused by longer patch cords. The rules for determining these lengths are covered later in this book. A MUTO may be appropriate for open plan or open office use or for temporary project offices, temporary repair/restitution or large retail spaces where connections have to be made between points-of-sale and structured cabling outlets placed in overhead gantries.

An 'interconnect' method of connection means connecting the TO to a patch panel via the horizontal cable and then connecting the front of that panel to the active equipment by way of a patch cord. This is shown in Fig. 3.9.

The interconnect method is the simplest and cheapest method of effecting a horizontal structured cabling system but at the loss of some network flexibility.

Cross-connect means having two patch panels at the end of the horizontal cabling. One is connected to the horizontal cabling and the other is connected to the active equipment. The appropriate con-

Floor distributor

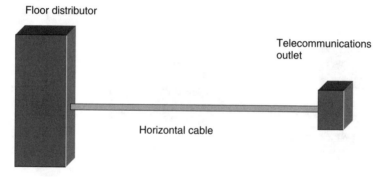

Telecommunications
outlet

Horizontal cable

**Fig. 3.9** Interconnect model.

Floor distributor

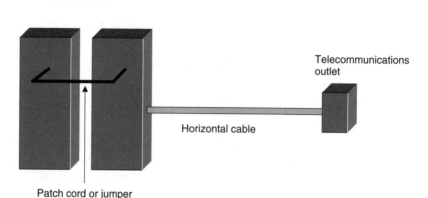

Telecommunications
outlet

Horizontal cable

Patch cord or jumper

**Fig. 3.10** Cross-connect model.

nection is then made by patching between the two patch panels. This is shown in Fig. 3.10.

Cross-connect gives the greatest flexibility by having one set of patching dedicated to the cabling and one set dedicated to the equipment. The downside is that there will be twice the number of patch panels required, consuming twice the space and doubling the amount of crosstalk and possibly return loss within the system.

For very large installations it may be impossible to get the active equipment and the patching equipment close enough together to get

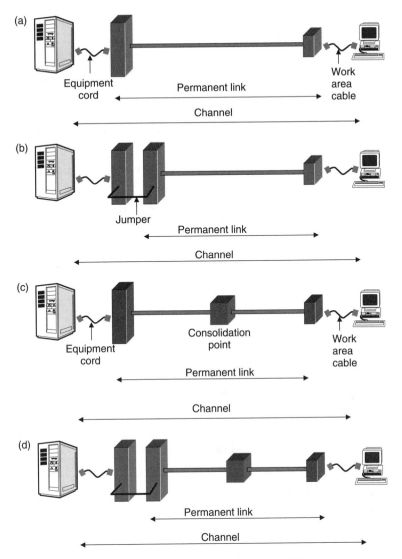

**Fig. 3.11** Horizontal cabling models: (a) interconnect-TO, (b) cross-connect-TO, (c) interconnect-CP-TO, (d) cross-connect-CP-TO.

away with just interconnect. For the more sophisticated intelligent patching systems that monitor what they are connected to it is essential to have cross-connect, or double representation as it is sometimes called. Intelligent patching systems usually work by using

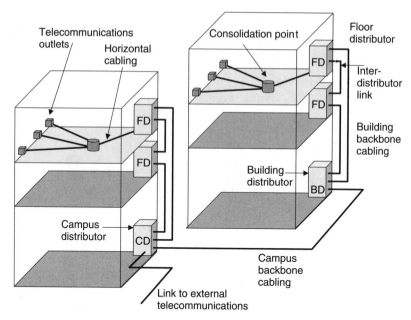

**Fig. 3.12** Generic campus cabling system.

special patch cords that contain extra connections that are used to identify what is connected to what. This is only possible with cross-connect.

We can see from the above that four combinations are possible:

- Interconnect-TO.
- Cross-connect-TO.
- Interconnect-CP-TO.
- Cross-connect-CP-TO.

All four possibilities are shown in Fig. 3.11.

Figure 3.12 gives a three-dimensional view of a campus cabling project showing virtually all possible combinations.

Most installations today have active equipment in the floor distributor (FD); a typical installation might consist of Cat5e copper cable to the desk running 10BASE-T with an Ethernet switch in the FD. The switch would be most likely to connect to an optical fibre backbone using a protocol such as 100BASE-FX. It is possible, however, to patch the horizontal cabling directly through to the backbone cabling, either on copper or optical cable. Once again care would have to taken that the communications protocols planned will be able to cope with the resultant attenuation and delay problems this will bring.

# 4

# Building blocks of structured cabling systems

## 4.1   Introduction

Structured cabling systems must be constructed from a family of building blocks or 'functional elements'. The exact detail of these elements will differ from supplier to supplier but they all share an agreed and standardised function in the design of the cabling system. The functional elements can be divided into cable and connecting hardware. Table 4.1 lists all the allowable functional elements in ISO

Table 4.1 Functional elements

| ISO 11801 2nd edition | TIA/EIA-568-B |
| --- | --- |
| Campus distributor (CD) | Main cross-connect |
| Campus backbone cable | Backbone cable |
| Building distributor (BD) | Intermediate cross-connect |
| Building backbone cable | Backbone cable |
| Floor distributor (FD) | Horizontal cross-connect |
| Horizontal cable | Horizontal cable |
| Consolidation point (CP) | Consolidation point (CP) |
| Consolidation point cable | Horizontal cable |
| Multi user telecommunications outlet (MUTO) | Multi user telecommunications outlet (MUTO) |
| Telecommunications outlet (TO) | Telecommunications outlet (TO) |

11801, EN 50173 and TIA/EIA-568-B. The terminology for the ISO 11801 and EN 50174 standards are identical but they differ from the TIA standard in that ISO 11801 refers to distributors whilst the TIA/EIA-568-B standard refers to the same items as cross-connects. Hence ISO 11801 may refer to a FD whereas TIA/EIA-568-B will call it a horizontal cross-connect.

## 4.2    Copper cables

In the ISO 11801 scheme, cables can appear in any of three locations:

- Horizontal cabling.
- Building backbone cabling.
- Campus backbone cabling.

If a CP is used then the cable from the CP to the TO may be referred to as the CP cable. Centralised optical architecture (COA) may combine any of the above three cabling subsystems.

### 4.2.1    Categories and characteristic impedance

One of the most important ways of describing the electrical performance of a copper cable is by its characteristic impedance. Unlike impedance or resistance, the characteristic impedance is not length related, although it uses the same units of measurement, namely the ohm ($\Omega$) and has the symbol $z_0$. In any electrical system it is essential that the output characteristic impedance of the transmission equipment is the same as the characteristic impedance of the cable system, which also has to be the same as the input characteristic impedance of the receiving equipment. If any of the elements in the transmission system have a different characteristic impedance then there will be electrical energy reflected back towards the source. In a power cable transmission system this could have disastrous effects upon the transmitter. In a communication system it is equivalent to an out-of-phase signal reflected back to the transmitter which will interfere and corrupt the data signal. In the field of structured cabling the amount of energy reflected back is called return loss, and it can be one of the most troublesome factors to overcome in an installation, especially for the higher frequency Category 6 cable systems.

Edition 1 and 1.2 of ISO 11801 allowed for the following types of cable by class and characteristic impedance:

* $100\,\Omega$, Category 3, 4 and 5.
* $120\,\Omega$, Category 3, 4 and 5.
* $150\,\Omega$.

It has to be said that 99.9% of all structured cabling installed in the world today is based on the $100\,\Omega$ standard. France Telecom generated the $120\,\Omega$ standard and 150-$\Omega$ refers to the characteristic impedance of the IBM Structured Cabling System, the first real universal cabling system.

ISO 11801 2nd edition retires Category 3 and 4 for both 100-$\Omega$ and 120-$\Omega$ variants and also the 150-$\Omega$ cabling disappears. This implies that 100- and 120-$\Omega$ Category 5, 6 and 7 cables are the only ones allowed. Cabling of 120-$\Omega$ is such a minor segment of the market that for the remainder of this book the assumption will be made that all structured cabling is based on 100-$\Omega$ characteristic impedance. Although ISO 11801 2nd edition gives a lot of cable detail, it refers to IEC 61156 as the overall cable standard, and this allows cable development to carry on outside the scope of this mainly systems-based standard. The equivalent cable standard referred from EN 50174 is EN 50288. TIA/EIA-568-B contains all its cable specifications within its section B-2, with Category 6 information published as Amendment 1 to section B-2. There is no equivalent to Category 7/Class E in the American standards.

Although Category 3 seems to have disappeared it is bound still to be required for backbone telephone cabling in large pair counts such as 100, 200, 300 pair and so on. This is still the most cost-effective way to transport PABX telephony in the backbone cabling, even if it has to go via Category 5 to the desk. In the UK there has always been a tendency to specify the British Telecom standard cable CW1308 for this job but the correct generic specification is Category 3. Category 3 can still be specified but the detail will be found in ISO 11801 1st edition or IEC 61156-2.

To summarise so far, the copper cables allowed in the system are Categories 5, 6 or 7, and will nearly always be based on 100-$\Omega$ characteristic impedance. Custom and practice still give an economic place for high pair-count backbone telephone (Category 3) cabling. This is not specified in ISO 11801 2nd edition but can still be specified for a cabling system by invoking ISO 11801 1st edition and/or IEC 61156.

## 4.2.2  Copper cable constructions

The cable construction allowed is twisted pairs or quads. At least eight conductors have to be in the cable, i.e. as in 'four-pairs'. Twisted pairs are by far the most common construction and are ideal for the balanced cable transmission technique used by modern LANs. Quads also give good quality balanced transmission but are only seen in France, to go along with the rest of the France Telecom 120-$\Omega$ cable specification. Figures 4.1 and 4.2 show twisted pair and quad

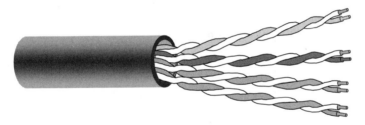

**Fig. 4.1** Unscreened twisted pair cable, UTP.

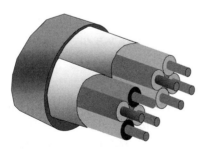

**Fig. 4.2** Quad cable.

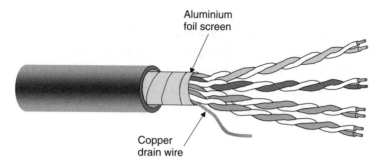

**Fig. 4.3** Foil twisted pair cable, FTP.

cable constructions. For the rest of this book we will presume that all copper cable constructions are based on four twisted pairs unless stated otherwise.

Copper cables can be either screened or unscreened. Screening, often referred to as shielding in American documents, means putting layers of metallic elements around the conductors to keep electrical interference out of the cable and to stop any electromagnetic signals leaving the cable. Screening the cable costs more and it makes the cables larger and more difficult to terminate. However, many people feel that some level of screening is essential for reliable network per-formance. In Germany around 98% of the data cable market is for screened product. In France it is about 60%. For the rest of the world the figure is nearer 10% screened and 90% unscreened. Making a choice between screened and unscreened cable is considered later in this book.

Unscreened cable, usually referred to as UTP (unscreened twisted pairs), consists of the four twisted pairs with an outer sheath or jacket to hold them together. There are several ways of screening a cable however:

• Putting an aluminium foil around the four pairs, under the sheath, along with a tinned copper wire known as the drain wire, which is there to facilitate a good grounding connection. This construc-tion is usually known as FTP, foil screened twisted pairs, see Fig. 4.3.

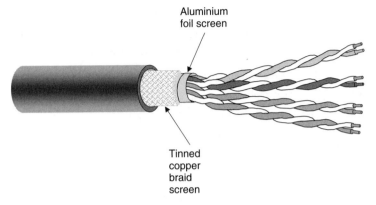

Aluminium
foil screen

Tinned
copper
braid
screen

**Fig. 4.4** Foil and braid screened cable, S-FTP.

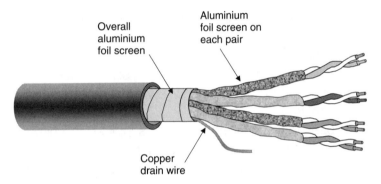

Overall
aluminium
foil screen

Aluminium
foil screen on
each pair

Copper
drain wire

**Fig. 4.5** Individually screened pair cable, STP/PIMF.

- Putting an aluminium foil and a tinned copper braid around the pairs. This is known as S-FTP, see Fig. 4.4.
- Putting an aluminium foil around each pair. This usually also involves an aluminium foil and/or a copper braid around all four pairs as well. This style is known as STP or PIMF (pairs in metal foil), see Fig. 4.5.

It must be noted that the terminology of screened cable can differ between different suppliers and it is essential to spell out specifically what is required when specifying screened cable to avoid product confusion.

ISO 11801 2nd edition does try to standardise the terminology with the following list, but it remains to be seen if this terminology is universally accepted, or understood, by all suppliers:

- U/UTP: completely unscreened.
- F/UTP: one foil screen around the whole cable.
- U/FTP: a foil screen around every pair.
- SF/UTP: a foil and a copper braid around the whole cable.
- S/FTP: A foil around every pair and a copper braid around the whole cable.

We can see that an 'F' (foil) or an 'S' (braid) before the forward slash denotes what goes around the whole cable and what comes after the slash denotes what goes around each pair, 'F' for foil or 'U' for 'nothing'.

TIA/EIA-568-B describes all screened cables as ScTP and states that 'an electrically continuous shield shall be applied over the core which shall consist of a metal laminated tape and with one or more drain wires'. The method of measuring screen effectiveness is usually the surface transfer impedance, which has units of milliohms per metre (m$\Omega$/m).

Copper cables will normally be supplied as units of four-pairs. Other constructions such as multiunits bundled together and bigger pair counts are allowed as long as they meet the same overall electrical specification, although for higher pair count cable the crosstalk specification has to be even higher to account for the close proximity of more conductors. Mixed hybrid units of copper and fibre cables or air blown fibre ducts are also allowed. Figure 4.6 shows such a bundled unit with an air blown fibre duct incorporated within it.

Category 6 cables usually require an extra constructional element to achieve the higher crosstalk isolation required for the 250 MHz Category 6 performance. This is achieved by inserting a plastic cross in between the four conductors to hold them apart mechanically. Figure 4.7 demonstrates this.

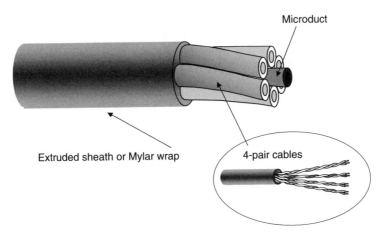

Microduct

Extruded sheath or Mylar wrap

4-pair cables

**Fig. 4.6** Copper cables and air blown fibre duct bundle.

**Fig. 4.7** Category 6 unscreened cable.

## 4.2.3　Indoor and outdoor grades of cable

The majority of structured cabling links are indoors, but cables will also need to run outside to link up buildings across a campus installation. Indoor cables have to be of low flammability. Outdoor cables must be weatherproof. Some cables are classified as universal or indoor/outdoor, meaning they are made from a sufficiently high grade material that meets fire regulations but is also weather resistant.

The cable design standards, such as ISO 11801 make little mention of fire performance, leaving it instead to more specific or local or national standards to take precedence.

In the USA cables are rated as outdoor, general purpose, riser or plenum, the latter three referring to positions in a building where a cable may be installed. Indoor cabling is covered under strict regu-

Table 4.2 American (US) copper cable ratings

| Cable marking | Type |
| --- | --- |
| MPP | Multipurpose plenum |
| CMP | Communications plenum |
| MPR | Multipurpose riser |
| CMR | Communications riser |
| MP, MPG | Multipurpose |
| CM, CMG | Communications |
| CMUC | Undercarpet |
| CMX | Communications, limited |

lation in the USA contained in the National Fire Prevention Association codes NFPA 70, with its National Electrical Codes (NECs). Table 4.2 lists American copper and optical cable ratings. This subject is covered in more detail later in this book.

In Europe fire safety standards are usually covered by IEC standards, but few of these are required by law in any European state. The main relevant IEC standards are:

- IEC 60332-1: *Flammability of a single vertical cable.*
- IEC 60332-3-24c: *Flammability of a bunch of cables.*
- IEC 60754: *Acid gas and halogen content.*
- IEC 61034: *Smoke evolution.*

However in the future, indoor cables installed within the EU will be covered by the Construction Product Directive and given Euroclasses of operation. A range of CENELEC standards is being written to cover mandatory Euroclass fire performance. The proposed Euroclasses are given in Table 4.3.

Outdoor cables have to be waterproof and if to be exposed to the sun they must also be ultraviolet (UV) light resistant. UV light will break down 'indoor' grade plastics within a few years.

Most external grade cables are sheathed in carbon-loaded black polyethylene to give a low cost level of protection against water and UV rays. Other methods include a polymer-coated aluminium foil tape under the sheath and then using petroleum gel to flood all the spaces

Table 4.3 Proposed European Construction Product Directive cable ratings

| Fire situation | Euroclass | Class of product |
|---|---|---|
| Fully developed fire in a | A | No contribution to a fire |
| room | B | Very limited contribution to a fire |
| Single burning item in a | C | Limited contribution to a fire |
| room | D | Acceptable contribution to a fire |
| Small fire effect | E | Acceptable reaction to a fire |
| | F | No requirement |

(the interstices) within the cable. Some optical cables avoid the use of gels by incorporating tapes and threads that expand upon contact with water. Outdoor cables are usually quite flammable owing to their construction and so are not usually allowed to enter a building by more than a few metres.

Cables that are going to be buried directly in the ground need extra layers of armouring. The two main armouring methods for copper cables are galvanised steel wire and corrugated steel tape.

## 4.3 Optical fibre and cables

### 4.3.1 Optical fibre

Optical fibres can be defined in several ways, for example, multimode and single mode, graded index and step index, etc. All optical fibres have the following in common: a core of a transmissive medium with a certain refractive index and a cladding of a material with a lower refractive index. It is this difference in refractive index that constrains the light mostly to remain within the fibre and not to escape from the sides. The core can be made of glass or plastic and the cladding can also be made of glass or plastic. The majority of optical fibres are all silica (i.e. glass), but there are fibres with a glass core and a plastic cladding or even composed entirely of plastic. The vast majority of optical fibres used in communications are made of all-silica. Plastic fibre has often been portrayed as the next generation low-cost alternative, but as yet it has not found many applications beyond the speciality market.

Four optical fibre types are defined in ISO 11801 2nd edition. They are all of all-silica construction and come in three multimode grades known as OM1, OM2 and OM3 and one singlemode grade, called OS1.

When rays of light enter the core of an optical fibre, as seen in Fig. 4.8, there are numerous routes they can follow. Some can go straight down the middle and some will be reflected from side to side. Each of these rays of lights can be considered as modes. A large-core fibre can support hundreds of these modes, hence the term, multimode.

This style of fibre, known as step index, suffers from a bandwidth limiting problem known as modal dispersion. This means that the rays of light, or modes, will spread out in time as the modes travelling the shortest route down the middle will get to the end before the modes taking the long way round, that is, the ones spending all their time bouncing from side to side. The answer to this is to vary the refractive index of the fibre core. The higher the refractive index of a medium the slower light will travel in it. Thus in data communications we use multimode fibres known as graded index, denoting that the value of

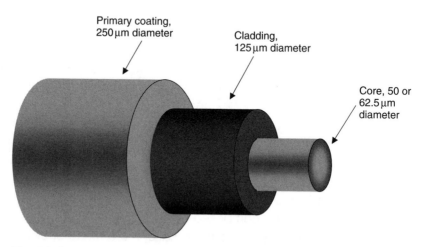

**Fig. 4.8** Optical fibre.

the refractive index is at a peak in the centre of the core and declines away the closer one gets to the cladding layer.

Single mode fibre, sometimes called monomode fibre, works in yet a different way. Single mode fibre is a step index fibre but the diameter of the core is much smaller than in a multimode fibre. A multimode fibre has a core diameter typically of 50 or 62.5 µm, or millionths of a metre, but a single mode fibre has a core diameter in the range of only 8–10 µm. This has the optical effect that for light rays above a certain wavelength only one mode will propagate down the core, hence the expression, single mode. Single mode fibre has a much larger bandwidth than multimode fibre because it does not suffer from modal dispersion. The wavelength at which the single mode operation comes into effect is called the cut-off wavelength, and is typically around 1250 nm (nanometres). Figure 4.9 shows the relative sizes of the three types of optical fibre used in data communications.

The first two fibres in Fig. 4.9 are specified in terms of the diameters of their core and cladding, respectively. The units are micrometres (microns), or millionths of a metre. Single mode fibre could also be described as 8/125. When the fibre is manufactured, a plastic coating is immediately applied to the glass cladding layer to

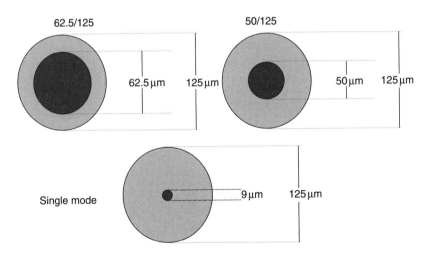

**Fig. 4.9** Optical fibres used in data communications.

raise the overall diameter up to 250 μm, or one quarter of a milli-metre. The fibre is then known as primary coated.

Single mode fibre is the very best and the cheapest. It has a huge bandwidth and a very low attenuation, yet it needs an expensive single mode laser to launch light into it. Traditionally multimode fibre is good enough for data communications with a capacity to send over 100 Mb/s over 2 km when driven by a relatively cheap light emit-ting diode, LED. The advent of gigabit transmission has shown up the age and shortcomings of multimode fibre. As LEDs cannot usually modulate above about 200 MHz then lasers have to be used for gigabit transmission.

A new class of laser has been invented called a VCSEL, or verti-cal cavity surface emitting laser. This device is currently only available for use on multimode fibre at 850 nm. Single mode VCSELs are on their way. VCSELs cannot transmit as far as conventional telecom-munications grade lasers but can transmit far enough for data com-munications use. The increase in LAN speed has already brought single mode fibre into the LAN campus backbone and if cheap single mode VCSELs become a reality then the economic rationale for mul-timode fibre will start to slip away as well as the technical reasons.

Choosing which fibre to use in a project will be discussed in the system design sections of this book. In this chapter we shall limit ourselves to defining the components.

A new 50/125 multimode fibre was introduced in 2001 to offer longer transmission distances for ten gigabit Ethernet; this is achieved by offering a much larger bandwidth than usual for multimode optical fibre. This fibre is known as OM3 in the new ISO 11801 2nd edition. It is also sometimes referred to as laser grade or laser enhanced. It offers a transition between the older styles of multimode fibre and single mode, but at a premium price!

In summary, OM1 can be 50/125 or 62.5/125 but in reality it is the standard 62.5/125 fibre that has been in use all through the 1990s. OM2 can also be 50/125 or 62.5/125 but it is most likely to mean the standard 50/125 also in use throughout the 1990s. OM3 is the new higher bandwidth 50/125 and OS1 is single mode.

Single mode fibre also covers a range of products, for example dis-persion shifted fibre and non-zero dispersion shifted fibre. For the

| Table 4.4 ISO 11801 2nd edition optical fibre classes | | | | |
|---|---|---|---|---|
| Fibre type | Core diameter (μm) | Minimum modal bandwidth (MHz km) | | |
| | | Overfilled launch bandwidth | | Effective modal bandwidth |
| | | 850 nm | 1300 nm | 850 nm |
| OM1 | 50 or 62.5 | 200 | 500 | N/s |
| OM2 | 50 or 62.5 | 500 | 500 | N/s |
| OM3 | 50 | 1500 | 500 | 2000 |
| OS1 | single mode | N/s | N/s | N/s |

N/s, not specified.

time being these are speciality products for the telecommunications market and will rarely be seen in premises cabling. The single mode fibre referenced as OS1 is the most common general purpose single mode fibre and is described more fully under ITU-T Recommendation G-652. Table 4.4 summarises the optical fibre performances defined in ISO 11801 2nd edition.

## 4.3.2   Optical cables

Optical fibres have to be protected during installation and during their service life, hence the need for a cable construction that is sufficient for that purpose. Optical cables can carry anything from just one to thousands of fibres and the constructions can vary enormously depending upon the environment in which they will be installed.

Optical cable designs fall into a number of categories but there are two principal groups, tight buffered and loose tube. Tight buffered fibres have a layer of mechanical protection in intimate contact with the fibre. Loose tube cables have tubes containing optical fibres that are loosely contained within them. There are some further variants of the latter such as slotted core and ribbon fibre. Tight buffered fibres, with their rugged nature, are preferred by many installers for their easier on-site termination. Loose tube cables give better mechanical protection to the fibres as the fibre is physically decoupled from most of the stress placed on the cable. Loose tube cables also give a better packing density of fibres for the same cross-sectional area and thus generally have a lower cost per cabled fibre. Other cabling options to be considered are various armouring and speciality sheathing variants.

## Tight buffered optical cables

Tight buffered optical fibre has extra layers of protection added directly onto the 250 µm primary coating to build up the fibre diameter to around 900 µm or 0.9-mm diameter. This style of fibre is viewed as being more rugged and easier to handle and best suited to on-site termination of the fibre, that is putting a connector directly onto the end of the fibre on site rather than splicing on a factory-terminated tail cable. Some versions are now available with a 600-µm diameter aimed at the small form factor (SFF) optical connector market.

Tight buffered fibre on its own may be used to make up short tail cables (i.e. a length of fibre with a connector on one end only), for use within patch panels or test equipment leads. Generally, however, the fibre needs be incorporated into a larger cable construction to be useful.

**Patch cord**  The simplest optical cable is a single tight buffered fibre contained within annular aramid yarns and held together within an outer sheath. The aramid yarns act as the strength member for the cable. The strength member is a concept common to all optical cables as the cable must be designed so that the optical fibre is subject to as little stress as possible.

The single fibre, tight buffered cable is made to a diameter of between 2.5 and 3.2 mm (typically 2.8 mm). All commonly used optical connectors are made with a back-end designed to accommodate such a size of cable. The back-shell of the connector body will terminate and grip onto the aramid yarn so that the yarn is subject to all of the tensile and other loads before the optical fibre.

Such a small and simple cable would only be used for patch cords and test leads. A variant is to extrude two such cables together to form a duplex structure, also known as zipcord, shotgun or figure-of-eight style. A yet more rugged construction is to extrude another

Cables used outdoors must be entirely adapted to the effects of the weather and waterlogged ducts or trenches. An outdoor cable must be waterproof, resistant to UV light and have a wide operating temperature range. The best materials for this application are polyethylene with petroleum gel filling. In addition, the polyethylene sheath is usually carbon loaded to improve UV resistance and this is why most external cables are black. Unfortunately polyethylene and petroleum gel are very flammable and most countries prohibit their use within buildings, insisting upon their termination to indoor grade cables within 15 m of entering the building. Failing this the cables must be entirely contained within a metal conduit.

Forming a transition joint between internal and external cables has never been seen as much of an issue with copper cables. Copper pairs are easily and quickly punched down onto blocks and in any case there is the need to break down the high pair-count external cable to the smaller pair-count riser cables. Added to this is the common requirement to insert overvoltage protection devices at the entrance facility.

Putting a transition joint into an optical cable is very labour intensive and expensive. The external cable route may itself only be a few tens of metres if buildings on the same campus are being connected and, of course, there is no need for overvoltage protection.

For this reason universal or indoor/outdoor grades of cable have become popular in the premises cabling industry. A universal cable has sufficient low flammability (and often zero halogen content) to meet national requirements yet maintains sufficient water and UV resistance to survive installation within flooded underground ducts. A higher performance material is required for the sheath to achieve these requirements yet the extra cost is easily absorbed within the lower labour costs of having only one cable to install on a site. On a typical industrial or academic park it is very common for cables to have to run tens of metres within a building and then tens of metres in between buildings. It is most cost effective in these instances to use just one common grade of universal cable.

To remove the need for water blocking petroleum gel, a newer method is to use water-swellable tapes and threads. Apart from the flammability issues with petroleum gel it is also unpopular with

Table 4.5 American (US) optical cable ratings

| Cable title | Marking | Test method |
| --- | --- | --- |
| Conductive optical fiber cable | OFC | General purpose UL 1581 |
| Non-conductive optical fiber cable | OFN | General purpose UL 1581 |
| Conductive riser | OFCR | Riser UL 1666 |
| Non-conductive riser | OFNR | Riser UL 1666 |
| Conductive plenum | OFCP | Plenum UL 910 |
| Non-conductive plenum | OFNP | Plenum UL 910 |

installers because of the mess and time taken to clean away all the gel before termination or splicing can take place. The threads and tapes swell considerably when in contact with water and this prevents water from flowing down the interstitial spaces within the cable.

There is an American (US) marking scheme for optical cables which dictates in which part of a building they may be installed. This is shown in Table 4.5.

## Loose tube optical cable

The other major family of optical cables is known as loose tube. The tube is typically 2–3 mm in diameter, gel filled and made of a material like PBT (polybutylene teraphthalate). Typically 12 fibres go in each tube but 96 fibres have been achieved.

The individual fibres are coloured to ease identification when splicing or terminating, but as the fibres are only 250-μm diameter, primary coated, the human eye could not readily differentiate between more than 12 colours. In cases where there are more than 12 fibres in a tube they are bunched together in groups by different coloured threads.

The main advantage of loose tube cabling is that the fibre is mechanically decoupled from the cable in the normal course of installation and operational use. As the tube is bent or stretched the fibre has some freedom to take up the position of least stress and strain.

Loose tube cables offer the highest possible packing density of fibres per unit area with hundreds or even thousands of optical fibres being contained in cables 20–30 mm in diameter. The majority of external cables are of the loose tube design owing to the superior strength they exhibit and their high fibre-packing density.

**Multiple loose tubes** For higher fibre counts and for maximum strength, the multiple loose tube cables are the most suitable. A typical design consists of a central strength member, which can be steel, aramid yarn or resin bonded glass (RBG) and around which is stranded six or more fibre tubes or fillers. The tubes are stranded around the central strength member to equalise the loading on the fibres. If the tubes were laid in straight, that is parallel to the central strength member, then when the cable was bent the fibres on the outside track would be under tension and the fibres on the inside track would be under compression.

The large central strength member makes the cable ideal for pulling into long underground cable ducts. A cable with six tubes of 12 fibres each packs a total of 72 fibres within a very small space. By varying the fibre count in the tubes, exactly the same cable carcass design can handle very different quantities of optical fibres. For lower fibre counts, some of the tubes may be replaced with simple plastic fillers just to keep the design circular. It is also necessary to have a method of identification for each fibre within the cable. Each fibre within each tube is coloured, as already discussed, and so each tube must also be coloured, or else just two tubes are coloured in what is known as a marker reference system.

Multiple loose tubes, with their high packing density, are very popular with PTTs (public telecom operators) but a high fibre count cable also needs a sophisticated cable management and termination regime, especially when bend-sensitive single mode fibres are being used.

Loose tube cables can be non-metallic, metallic and may also be armoured for more severe environments such as direct burial. Totally non-metallic optical cables are more popular in data communications applications whereas longer distance telecommunications applications are still content to use metallic construction optical cables.

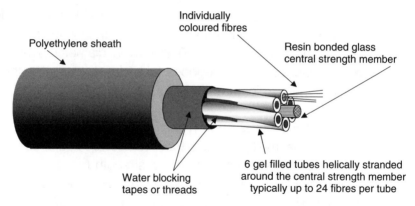

**Fig. 4.13** Non-metallic loose tube optical cable.

A typical non-metallic cable would contain a non-metallic central strength member such as RBG. Stranded around it could be six gel filled fibre tubes. The interstitial spaces between the tubes may also be flooded with gel if it is an outdoor cable under which case there would be a layer of paper tape to hold the whole thing together while it went through the next manufacturing process, which is extrusion of the final sheath. The final sheath would be polyethylene for an external grade cable.

If it was a universal cable then the interstitial gel filling would be replaced with water-swellable tapes and threads and the outer sheath would be low-flammability zero-halogen compound. Figure 4.13 illustrates a typical multiple loose tube, non-metallic duct cable.

**Single loose tube optical cable**    A low cost alternative to multiple loose tube designs is to have just one fibre tube in the middle of the cable with annular strength members around it and an overall sheath. Some manufacturers also refer to this style as unitube.

The tube is generally the same as in multiple loose tubes but may be slightly larger. The cable can be made in external duct style or universal grade. Unitube is not as strong as multiple loose tube with its big central strength member but unitube is probably the lowest cost design available and is suitable for shorter campus links. Figure 4.14 demonstrates the unitube style.

Layer of aramid yarn strength
members and water swellable threads

Outer sheath
of LSF/0H material

Up to 48 primary coated
fibres (in bundles of 12) in a gel filled tube

**Fig. 4.14** Unitube optical cable.

## *Armouring styles*

If a cable is not being installed in the relatively benign environment of an office or even an external cable duct then it probably needs to be armoured. Armouring gives the cable much greater compressive strength and prevents attacks from rodents and insects.

**Steel wire armour**   Steel wire armour, or SWA, is the strongest of the armouring options. It involves stranding galvanised steel wires around the cable and then extruding over it a final sheath. Figure 4.15 shows a cable typical of heavy duty use such as direct burial in the ground in a cable trench.

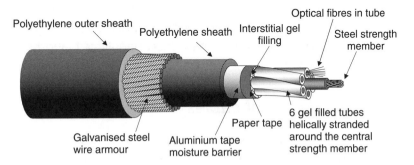

Polyethylene outer sheath

Polyethylene sheath

Interstitial gel
filling

Optical fibres in tube

Steel strength
member

Galvanised steel
wire armour

Aluminium tape
moisture barrier

Paper tape

6 gel filled tubes
helically stranded
around the central
strength member

**Fig. 4.15** Steel wire armoured optical cable.

**Corrugated steel tape**   A lighter, more flexible and lower cost alternative to steel wire is corrugated steel tape. The tape appears quite thin when seen in its original form, but when corrugated and over-sheathed with polyethylene it provides a very strong layer of protection. This construction of cable is especially popular to resist rodent attack. Figure 4.16 demonstrates this style.

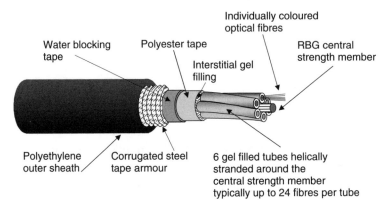

**Fig. 4.16** Corrugated steel tape armoured optical cable.

**Non-metallic armour**   The options described above all add metal to the cable. Some non-metallic armouring options are:

*   *Glass yarn.* A layer of glass yarn applied under the outer sheath gives a useful degree of added strength to the cable. It is also very difficult and unpleasant for rodents to gnaw through the cable.
*   *Nylon.* Nylon is very hard and slippery and when added over a sheath of high-density polyethylene makes the cable very tough and almost too slippery for small rodents to get a grip onto with their teeth. Nylon has the added benefit of being oil and chemical resistant.

## Special optical cables

There are many projects and applications where special designs of cables and even specialised fibres are required. Some of the applications and design options are summarised here.

- *Copper and fibre composite cables.* Copper wires and optical fibres can be mixed in the same cable. The copper wires can be used for carrying power to remote equipment or to carry low speed data or switching signals. One design places optical fibres in the centre of the cable and puts a layer of twisted copper pairs around the outside.

- *Lead sheathed cable.* In oil refineries it is often a requirement that cables carrying process-critical information must be lead sheathed. This is because hot crude oil will eventually dissolve any kind of plastic, no matter how exotic its make-up. Lead is totally impervious to oil and will guarantee the signal integrity forever.

- *Radiation resistant fibre.* Optical fibre goes dark in the presence of ionising radiation, i.e. the attenuation goes up to unacceptable levels and the circuit will cease to work. The effect is more pronounced in multimode fibres. Some military and nuclear power/ processing installations therefore specify radiation resistant fibre. Another option is, of course, to use the lead-sheathed cable.

- *Oil/chemical resistant cables.* There are a wide variety of materials that have a greater resistance to solvents than can be obtained from day-to-day materials such as PVC and polyethylene. Nylon provides one of the cheapest means of obtaining oil and chemical resistance. Applications in places such as coal mines and oil rigs need cables that are resistant to solvents ranging from sea water to hydraulic fluid. A resistance to ozone and UV light may also be needed, as is a requirement to maintain flexibility across a wide range of temperatures. A useful specification that can be used here is the British Naval Engineering Specification DEF-STAN 61-12 part 31.

- *Field deployable cables.* In some cases it is necessary to roll out an optical cable, use it and then roll it up again and take it away. Conventional optical cables, their connectors and even the cable drums they come on are all totally unsuitable for this application. Cables, connectors and the special drums that can be used for this requirement are often called field deployable. The cables involved have to be very strong and flexible over a wide temperature range and the most popular construction for this is four tight-buffered fibres within a bed of aramid yarn all contained within a polyurethane outer sheath. Special connectors, called expanded beam connectors, are also required.

## *Blown fibre*

Blown fibre, or air blown fibre, is not to be confused with air blown cables. An air blown cable is a relatively conventional optical or copper cable that is assisted in its passage through a cable duct by the action of air pumped alongside it. Blown fibre is similar in concept but consists of specially coated fibres or groups of fibres blown into small plastic ducts.

Blown fibre was pioneered by British Telecom in the 1980s and is still extensively used by BT in their local network. BT has licenced several manufacturers around the world to produce their own versions.

The purpose of blown fibre is to reduce initial installation costs. Rather than installing large quantities of dark fibre, that is, spare unconnected fibre for future use, blown fibre allows for the installation of low cost plastic ducting that can have optical fibre blown into it at a future date as and when it is needed. Blown fibre clearly becomes more economic the larger the project. Small installations requiring a single eight-fibre backbone, for example, may be better off with a conventional optical cable, but a large site with potentially thousands of fibre links worth hundreds of thousands of pounds can save over 65% of initial cabling costs with blown fibre. The final costs, once the blown fibre is added, are roughly the same as conventional cable or even slightly higher. But the main point is the saving of large amounts of capital in the early years of a project. To summarise, the advantages of blown fibre are:

- The ability to defer costs into the future.
- The ability to blow out old fibre and re-use the ducts.
- The ability to defer difficult decisions about fibre type into the future, e.g. single mode fibre, laser-launch optimised 50/125, etc.
- The ability to repair the ducts more easily than conventional optical cable.
- The ability to add fibre without any disruption to the office working environment.

The disadvantages are:

- Limited number of suppliers.
- No cost advantage for small projects.
- The need to record and protect the ducts so they can be used in the future.

**Blown fibre ducts**   The blown fibre ducts are plastic tubes around 5–8 mm in diameter. They are supplied singly or in groups, typically four or seven or even more and packaged within an outer sheath. Waterproof and armoured versions are available for outdoor use and flame retardant versions for indoor use.

The build-up of static electricity is the biggest problem encountered by blown fibre ducts. A plastic coated fibre blown against the inside of a plastic tube would end up sticking to it owing to static after a few tens of metres. The duct therefore has an inner lining of carbon-loaded polyethylene that is sufficiently conductive to leak away static electricity.

The ducts are easily joined together, for example from outdoor grade to indoor grade, by the use of push-fit pneumatic connectors. Figure 4.17 shows a group of blown fibre ducts.

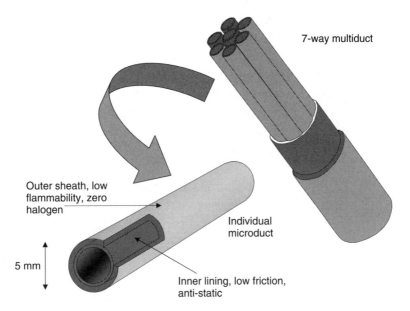

7-way multiduct

Outer sheath, low flammability, zero halogen

Individual microduct

5 mm

Inner lining, low friction, anti-static

**Fig. 4.17** Blown fibre ducts.

**Blowable fibre**   The optical fibre used is standard primary-coated fibre. Applied to it is a low-friction, anti-static coating to raise its diameter up to around 500 μm or 0.5 mm. Alternatively a bundle of fibres, such as four or seven are oversheathed with a low friction material. Optical ribbons can also be blown into ducts. The coating material is generally a UV cured acrylate containing PTFE chips or tiny glass beads and this combination gives the desired qualities of low static build-up, low friction and a high aerodynamic drag factor.

**Blowing equipment**   Either clean dry air or any other clean dry gas, such as nitrogen, can be used. Compressed air or gas can be supplied in bottles or compressed air can be generated locally by use of a compressor. The compressor does not have to be particularly special but it must have the capability to clean and dry the air. Flow rates vary but obviously a larger compressor is required to blow fibre 2 km rather than 50 m.

A machine is needed to guide the fibres into the duct and allow the air to flow over the surface of the fibre and pick it up. The blowing machine is based on a simple tractor feed mechanism that pushes the fibres into the ducting. The blowing machine needs to be able to control fibre speed, air flow, keep count of the fibre quantity blown in and be able to detect fibre blockages and to stop automatically.

A major difference with blown fibre ducts compared to optical cable is the need to pressure test the ducting before the fibre is blown in. This is to ensure that the duct has not been crushed, kinked or punctured during installation. The duct is similar in size and strength to Category 5 copper cable but it still needs to be installed with some care.

The pressure test consists of blowing a steel ball from one end of the duct to the other. At the remote end is a special valve that catches the ball. If the ball arrives one can safely assume that the duct is not damaged. To check the pressure integrity, however, the ball closes off a valve when it arrives at the far end which then allows the air pressure to rise in the duct, if the compressor is still turned on. The equipment allows the pressure to rise to its maximum working pressure of 10 bar, and if the pressure holds then it can be concluded that the duct has no punctures or tears in it. The duct is then depressurised and sealed at both ends. This is important in order to prevent dust, water and insects from entering into the duct. The sealed duct is then ready to have fibre blown into it at any time in the future, although, if a considerable time has elapsed, it would be wise to do another pressure check before attempting to blow fibre.

**Blown fibre capabilities**   Blown fibre bundles are optimised for the long straight distances that may be encountered in external routes. As much as 2 km may be blown in one go and one method even allows fibre to be picked up and blown into the next section.

For internal cable routes, like those usually encountered in LANs, numerous tight bends that need to be negotiated are more of a problem than achieving long route distances. Individual blown fibres are better at taking numerous bends and one method allows up to 12 optical fibres to be blown along a route containing three hundred 25-mm bends.

The individual fibre blowing method can blow fibre 500 m through the 5-mm duct and over 1000 m through the 8-mm duct. Vertical rises of at least 300 m are also possible.

With both ends disconnected, and using the same compressor, the optical fibres can be easily blown out and the duct reused. Additional fibres cannot be added to an already populated duct as they would intertwine and be unable to progress. When blown fibre is used for backbone cabling it is common to use a seven tube bundle, which is potentially an 84-fibre cable, and to blow in four or eight fibres for immediate use, leaving the capacity of the other six tubes available for future expansion.

## 4.4   Connecting hardware

### 4.4.1   Functional elements

As has been shown, the cable can appear in a number of locations, such as the horizontal or backbone. The cable is terminated by connecting hardware which can be either copper or optical. Where the cable arrives at the work area the outlet is known as the TO.

At the other end of the horizontal cabling is the FD (horizontal cross-connect) which also acts as the connection between the horizontal cabling and the building backbone cabling, if there is any.

Between the building backbone and the campus cabling is the building distributor (intermediate cross-connect). At the top of the hierarchy is the campus distributor (main cross-connect).

Within the horizontal cabling we may also find a CP or a MUTO.

As already stated, the connecting hardware within these distributors and outlets can be either copper or optical and is essentially a range of connecting blocks or plug and socket combinations.

## 4.4.2   Copper connecting hardware

The copper connecting hardware is either based on 8-pin modular connector plugs and sockets, known as RJ-45s (although they are not referred to as such in the standards), or strips of a connector style known as insulation displacement connectors (IDCs). The TO and MUTO are always plug and socket arrangements but the distributors and CP can be plug and socket or strips of IDCs.

The RJ-45 8-pin modular connector plug and socket gives the greatest possible flexibility in terms of user patching. The IDC strips are lower cost, have a greater packing density but need specialised technicians on hand to make any major change.

## *RJ-45 Modular 8-pin connector*

Structured cabling systems for LANs have now been standardised to use an 8-pin modular connector commonly known as an RJ-45. RJ-45 is a USOC (universal service order code) designation for specific applications and/or manufacturers, and the correct generic title for this connector is IEC 60603-7, although RJ-45 has come to be the commonly accepted term. All three system design standards refer to IEC 60603 for the connector details and do not mention RJ-45, but the term is commonly accepted by manufacturers, installers and end users. Figure 4.18 shows a typical RJ-45 socket.

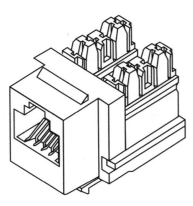

**Fig. 4.18** Typical RJ-45 socket.

The older versions of ISO 11801 also recognised the IBM cabling System 150Ω connector. This is not included in ISO 11801 2nd edition.

The RJ-45 has eight pins so it can terminate a four-pair cable. There are other 'RJ' styles such as the RJ11, which has six pins and is commonly used in telephony, and the RJ21x, which is a 50-pin, 25-pair connector.

The RJ-45 pins still have to be connected to the four-pair cable. This may be done by an IDC, which terminates the cable. The IDC is hard wired to the pins of the RJ-45 or it may be connected via a small printed circuit board (PCB) which helps to cancel out some of the crosstalk inherent in the connector design. The RJ-45 may use any style of insulation displacement connectors and can come in unscreened or screened format.

The choice of which conductor in the cable is connected to which pin on the front of the RJ-45 is decreed in TIA/EIA 568B and is referred to by its American (US) designation, 568A and 568B. This has nothing to do with the names of the overall cabling standard once known as 568A and now 568B. ISO 11801 does not specify which pairs on the cable should be connected to which pins on the RJ-45.

The two methods are often just called 'A' or 'B' wiring. In essence, pairs one and three are swapped over as shown in Fig. 4.19. There is little, if any, technical difference between the two wiring standards but it is essential to remain consistent within the same installation.

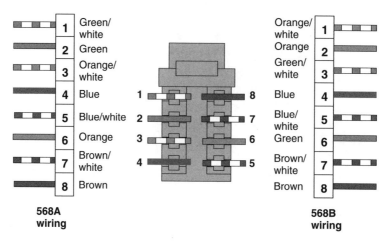

**568A wiring**

| Pin | Colour |
|---|---|
| 1 | Green/white |
| 2 | Green |
| 3 | Orange/white |
| 4 | Blue |
| 5 | Blue/white |
| 6 | Orange |
| 7 | Brown/white |
| 8 | Brown |

**568B wiring**

| Pin | Colour |
|---|---|
| 1 | Orange/white |
| 2 | Orange |
| 3 | Green/white |
| 4 | Blue |
| 5 | Blue/white |
| 6 | Green |
| 7 | Brown/white |
| 8 | Brown |

**Fig. 4.19** 'A' and 'B' wiring.

The RJ-45 has a Category of operation, just like the cable. ISO 11801 2nd edition recognises Categories 5, 6 and 7.

It is essential that all components in the cable system are designed to the same category of performance. The overall system performance of the cabling link is only as good as the lowest common denominator. For example, a cable system may have Category 5 cable terminated by Category 5 connectors, but if it is connected into the LAN equipment by Category 3 patch leads then the overall channel performance will only be Category 3/Class C.

**Category 7 connectors**   The RJ-45 was only designed to cope with 3-kHz telephone signals and it is unlikely to exceed the demands of Category 6 with its requirements for a 250-MHz frequency range. Category 7 requires 600-MHz performance. ISO has approved a connector, designed by Nexans, which has Category 7 performance yet is backwardly compatible with RJ-45 connectors. It achieves this by having eight pins in the socket as in a conventional RJ-45, and four more, arranged as two pairs in each corner. When in RJ-45 mode, that is Category 6 and below, only the eight pins on the top are engaged. When the socket detects that Category 7 operation is required (by detecting a lug on the side of a Category 7 plug), the four middle pins on the top deck are disconnected and the two pins in each corner are engaged, thus giving sufficient physical separation of the four pairs so that Category 7 performance is achieved.

As an alternative design, the ISO committee has approved a connector design from The Siemon Company for Category 7.

Category 7 only comes as a screened product.

## Insulation displacement connectors

An IDC is essentially a piece of metal with a 'V' shape cut into it. The narrowest point of the 'V' will have a distance slightly less than the diameter of the copper conductor it expects to encounter. When a plastic insulated wire is pushed down into the 'V' (the 'punchdown') with a special tool, the sharp inside edge of the 'V' cuts through the insulation and makes contact with the copper conductor inside. This saves the laborious task of manually stripping off the insulation and then screwing down or soldering the contact. The most popular styles of IDC are the '110' ('one-ten') developed by AT&T and the LSA from Krone. Also in use are the '66' and BIX connectors from Siemon and Nordx/CDT, respectively. Figure 4.20 shows a 110 IDC block.

Strips of IDCs, typically 50 or 25 pairs in a row are often called cross-connect strips. The IDC in a cross-connect usually works by having a 'lower deck' which is just a row of 50 IDCs. Onto this is placed a connector block that is made for two, three, four or five pairs. The exit wire or jumper can be punched down onto this upper deck formed by the addition of the connector block. There are special connectors that can fit directly onto the top of the IDC to provide an easily removable jumper lead.

The cross-connect is also used to break down high pair-count backbone cables into smaller pair-count cables. For example, a 100-pair backbone cable may need to be broken down into four-pair cables for onward routing to the desk position. In such a case 25 pairs at a time would be laid down onto the lower deck of the IDC cross-connect. A four-pair connector block is pushed down onto the first four terminated pairs of the backbone cable and makes electrical contact with them. The four-pair exit cable is then punched down onto the top deck of the connector block. In telephony terms such

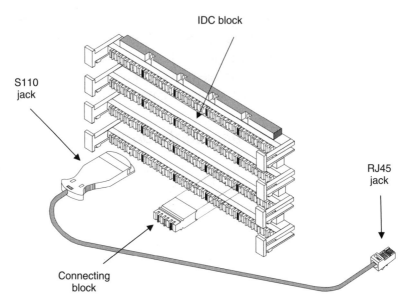

**Fig. 4.20** 110 IDC block.

an arrangement would be called an intermediate distribution frame (IDF).

The IDC must have a Category of operation, just like the cable. ISO 11801 2nd edition recognises Categories 5, 6 and 7 but in reality most telephone distribution systems use IDC-based distribution frames with not more than Category 3 performance.

## *Patch panels*

For easy user patching the RJ-45 sockets are organised in a patch panel. Patch panels are 19 inches wide (483 mm) to fit into the industry standard equipment racks. Perversely the units for all three dimensions of a patch panel are different. The width is always 19 inches, the depth is in millimetres and the height is in 'U'. One U is 44 mm. The standardised notation of 'U' allows designers to calculate quickly how much equipment can be loaded into an equipment rack. So, for example, a 30 U rack could accommodate 30 × 1-U patch panels. A 1-U patchpanel, see Fig. 4.21, will usually have 12, 16 or 24 RJ-45

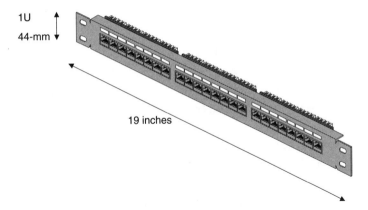

1U

44-mm

19 inches

**Fig. 4.21** RJ-45 patch panel.

sockets on the front. An IDC on the back will terminate each cable and supply the connection through to the pins of the RJ-45 on the front. There may or may not be extra cable management supplied on the back. The patch panel may be screened or unscreened and wired in either 568A or 568B format. A patch panel may be larger than 1 U. A 2-U panel can typically terminate up to 48 connectors and there are even 4-U panels for 96 connectors.

The RJ-45 patch panel has a Category of operation, just like the cable. ISO 11801 2nd edition recognises Categories 5, 6 and 7. Some patch panels, especially for Category 7 operation, are just individual RJ-45 sockets housed within an appropriate 1-U metalwork component.

## Intelligent/managed patch panels

The latest generation of patch panels offer a level of asset management and connectivity information by sensing what they are connected to.

One way of doing this is to have a special patch cord with at least one extra wire in it and an RJ-45 with 10 pins rather than the usual eight. This gives enough extra connectivity through this special patch cord for one patch panel to know which other patch panel it is connected to. This method of intelligent patching can only work when the cabling is used in the cross-connect or double representation format.

Another method uses a more conventional patch cord but has a spring loaded 'detector' pin above each RJ-45 socket. The panel thus detects when a connection has been made and assumes that the next detected connection is the other end of the same patch cord.

To make use of the information requires a 'scanner'. The scanner is a piece of electronics that resides in each equipment rack and is connected to each of the managed patch panels by a ribbon cable or similar. The scanner interrogates each of the panels in turn to derive the connectivity information, which can then be relayed back to the supervisory station and/or added to the overall connectivity database of the enterprise.

All this extra connectivity and management information will, of course, come at a price, but in larger installations it may well lower the overall IT management costs and cable-related downtime costs.

Intelligent patch panels still need to be rated in terms of Category 5, 6 or 7 and screened or unscreened. Such patch panels must also be failsafe, meaning that if the electronic scanning function fails, the system still operates as a conventional passive cabling system.

Manufacturers have experimented with completely software controlled remote patching and switching. This was achieved up to about Category 3 and even 'old' Category 5 levels of performance, but the costs of maintaining Cat5e, Category 6 or 7 electrical performance through an electronic switch have so far proved to be uneconomic.

## *Patch cords*

The final element of the copper connectivity is the patch cord. A patch cord is a short length of stranded conductor cable with (usually) an RJ45 plug on each end. It links patch panels, cross-connects and the active LAN equipment at each end of the cabling channel. The exact terminology depends upon where the patch cord is used, for example:

- Work area cords: connect TOs to terminal equipment.
- Equipment cord: connects IT equipment to the generic cabling at distributors.
- Patch cord: connectorised cable making connections at a patch panel.
- Jumper: an assembly of twisted pairs without connectors used at a cross-connect.

## *Other components*

The following components are not described as functional elements in ISO 11801 2nd edition but they are often essential elements in a practical cabling system and so are briefly described here.

**Baluns**    A balun (balanced–unbalanced) is an impedance matching device that can be used to connect an unbalanced coaxial cable of impedance between 50 and 125-$\Omega$ and a balanced 100-$\Omega$ twisted pair cable. The balun has to be specifically made for the application so that one might be described as 75-$\Omega$ to 100-$\Omega$ for example. The 100-$\Omega$ side would be a standard 8-pin RJ45 and the 75-$\Omega$ side would be the appropriate BNC coax connector. Baluns may be screened or unscreened and sometimes there can be different pins connected within the RJ45, so they are best bought from the same supplier, in pairs. The electronic transmission equipment, which may have been made to work on 93-$\Omega$ coax for example, should be unaware that it is really communicating over 100-$\Omega$ four-pair cable, if the correct baluns have been used.

Video baluns can be baseband (8-MHz composite video, e.g. a CCTV camera), broadband (500–700-MHz CATV) or RGB, which puts red, green and blue signals on three separate pairs within the same cable. This system is used for high quality CAD/CAM graphics and dealer-desk systems.

**Adapters**    Devices such as telephones and ISDN equipment can work over structured cabling but usually cannot plug directly into the RJ45 TO. Some form of adapter is often required, which is often dictated by the style of PABX in use. The manufacturer of the PABX/ISDN equipment should be consulted to determine which style of adapter is required, for example secondary line adapter, PABX line adapter or PSTN line adapter.

It must be noted that the standards only allow for RJ45 copper cable connectors or optical connectors in the TO, nothing else is allowed. So any balun or adapter must be external to the RJ45. This allows the device to be removed at any time letting the cabling revert to its open-systems applications-independent role.

**Cable organisers**   Racks containing large numbers of patch panels can create unmanageable piles of patch cords cascading down the front. A 30-U rack could have 720 patch cords with an average 3-m length in front of it, that is over 2 km of cable! Cable managers and organisers should be inserted in between patch panels at a spacing of every 3 or 4-U to route the patch cords neatly away from the front of the patch field.

**Line protection devices**   Where external copper cables are used they should be protected from over-voltage and surge current conditions at the point where they enter the building. Lightning strikes in particular can cause damaging and dangerous voltage spikes to enter the building cabling, posing a threat to life and equipment.

Primary protectors should be placed across the incoming circuit and ground as close as possible to the place where the cable enters the building. For service entrances this may be the point of demarcation between the PTT cable and the user's own cabling. There are three main types of primary protector:

- Carbon block.
- Gas tube.
- Solid state.

All work on the principle that at a certain voltage the route to earth will break down and shunt the fault current to ground. There are also devices known as secondary protectors, which are often designed to defend against lower level, but perhaps more persistent fault or 'sneak currents'.

Any device attached to a high speed LAN cable must be designed so that it does not degrade the high performance potential of the cable.

## 4.4.3   Optical connecting hardware

The functional elements of distributors, cross-connects, CPs and TOs are exactly the same whether they are implemented in copper or optical components, but of course the optical components are very different from their copper counterparts.

## *Optical connectors*

Every optical fibre has to be terminated by a connector to allow it to plug into the active communications equipment or to join another connectorised fibre in an optical patch panel.

Most optical connectors are of the ferrule type, that is, they consist of a precision-made ceramic or polymer tube that has an internal diameter of a few micrometres more than the cladding of the fibre. In most cases this is approximately 125 µm. The rest of the body of the connector is designed to support and protect the ferrule and the fibre that goes into the back of it. Two of these connectors are designed to mate together through an adapter or uniter so that the polished end faces of the fibre cores are closely aligned, allowing the light to cross from one core to another with an acceptable loss *en route*.

There are many different types of connectors in circulation, most with the various termination options. Two of the earliest styles were known as SMA and Biconnic. In data communications and premises cabling the two styles that dominated the market in the 1990s were the ST (or ST2 as it is often referred) and the SC, shown in Figs. 4.22 and 4.23.

In telecommunications, where single mode fibre dominates, the two main connector styles used are the FC-PC and the SC.

Connectors are often described as multimode or single mode yet both are essentially tubes with 125-µm holes in the middle. The difference is in the tolerance with which the components are made. A

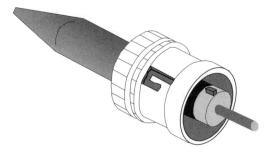

**Fig. 4.22** ST optical connector.

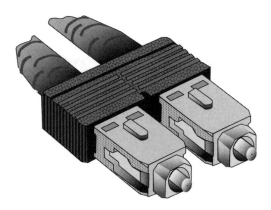

**Fig. 4.23** SC duplex optical connector.

multimode connector will typically have a hole size of 127 (+4, –0) µm and a single mode connector will have a hole size of 126 (+1, –0) µm. The fibre specification is 125 ± 2 µm. The ferrule hole size must, of course, be bigger than the fibre size or it would never fit into the ferrule, but whereas a 62.5/125 fibre could easily cope with a 4-µm misalignment, it would be a disastrous mismatch for an 8-µm single mode core.

Other connectors in circulation include the duplex FDDI MIC (media interface connector) and the similar IBM ESCON® (Enterprise System Connectivity Architecture) connector. ESCON connectors are made specifically for the IBM Enterprise Connectivity architecture for connection between mainframes and high-speed peripherals.

Optical connectors usually have to be recognised by an IEC standard number, but they usually start life when a manufacturer or group of manufacturers take an idea to the American FOCIS (Fiber Optic Connector Intermateability Standard) committee, TIA/FO 6.3.4. This committee will assess the viability and technical merits of the connector and the members of the committee will vote on its acceptance. If a connector is recognised by FOCIS nobody is obliged to use it, but other standards bodies such as ISO may then specify it as part of a LAN or cabling system standard. If a connector is seen to be achieving market acceptance, as well as giving some technical/economic benefit, electronic hardware manufacturers will start using it in their equipment.

The ST and SC connectors have dominated the LAN/premises cabling market for most of the 1990s, but a new generation of connectors is now making an appearance. Most of them fall under the heading of SFF (Small Form Factor). The concept of SFF is to make a multifibre connector of roughly the same footprint as a copper RJ-45 connector. Table 4.6 gives the current line-up of connectors in the FOCIS list and Table 4.7 gives the IEC listing of optical connectors.

ISO 11801 2nd edition, states that the optical connector at the TO shall be a duplex SC connector to IEC 60874-19-1. Optical connectors at other locations shall meet the optical, mechanical and en-

Table 4.6 TIA/EIA 604 FOCIS optical connectors

| Standard | Date | Title |
|---|---|---|
| TIA/EIA 604-1 | April 1996 | Biconic |
| TIA/EIA 604-2 | November 1997 | ST |
| TIA/EIA 604-3 | August 1997 | SC |
| TIA/EIA 604-4 | August 1997 | FC |
| TIA/EIA 604-5 | November 1999 | MPO |
| TIA/EIA 604-6 | March 1999 | Fiber Jack |
| TIA/EIA 604-7 | January 1999 | SG (VF-45) |
| TIA/EIA 604-10 | October 1999 | LC |
| TIA/EIA 604-12 | September 2000 | MT-RJ |

Note: The missing numbers, e.g. FOCIS 11, are for connector proposals that are still in discussion, in draft, or in ballot. For example the proposal for FOCIS 11 is the Siecor SCDC/SCQC connector.

Table 4.7 IEC standards for optical connectors

| Standard | Date | Title |
|---|---|---|
| IEC 60874-7 | 1993 | FC |
| IEC 60874-10 | 1992 | BFOC/2.5 (ST) |
| IEC 60874-10-1 | 1997 | BFOC/2.5 multimode |
| IEC 60874-10-2 | 1997 | BFOC/2.5 single mode |
| IEC 60874-10-3 | 1997 | BFOC/2.5 single mode and multimode |
| IEC 60874-14 | 1993 | SC |
| IEC 60874-14-1 | 1997 | SC/PC multimode |
| IEC 60874-14-2 | 1997 | SC/PC tuned single mode |
| IEC 60874-14-5 | 1997 | SC/PC untuned single mode |
| IEC 60874-14-6 | 1997 | SC APC $9^0$ untuned single mode |
| IEC 60874-14-7 | 1997 | SC APC $9^0$ tuned single mode |
| IEC 60874-14-9 | 1999 | SC APC $8^0$ tuned single mode |
| IEC 60874-14-10 | 1999 | SC APC $8^0$ untuned single mode |
| IEC 60874-16 | 1994 | MT |
| IEC 60874-19 | 1995 | SC duplex |
| IEC 60874-19-1 | 1999 | SC-PC duplex multimode patch |
| IEC 60874-19-2 | 1999 | SC duplex single mode adaptor |
| IEC 60874-19-3 | 1999 | SC duplex multimode adaptor |
| IEC 61754-2 | 1996 | BFOC/2.5 (ST) |
| IEC 61754-4 | 2000 | SC |
| IEC 61754-5 | 1996 | MT |
| IEC 61754-6 | 1997 | MU |
| IEC 61754-7 | 2000 | MPO |
| IEC 61754-13 | 1999 | FC-PC |
| IEC 61754-18 | draft | MT-RJ |
| IEC 61754-19 | draft | SG |
| IEC 61754-20 | draft | LC |

vironmental requirements stated in IEC 60874-19-1. This seems to be a clear indication that at all locations apart from at the TO, any IEC recognised optical connector is allowed. The minimum optical performance of the connector, see Table 4.8, is identical regardless of which connector is used.

The TIA/EIA 568B Standard differs in that it will allow the use of any FOCIS recognised connector but the optical performance and test methods must still conform to that Standard's requirements, and the connector must meet the requirements of the TIA FOCIS document. The SC duplex connector is given as an example in the

Table 4.8 ISO 11801 Optical connector performance

| | 850 nm multimode (dB) | 1300 nm multimode (dB) | 1310 nm single mode (dB) | 1550 nm single mode |
|---|---|---|---|---|
| Optical connector, insertion loss | 0.75 | 0.75 | 0.75 | 0.75 |
| Optical splice, insertion loss | 0.3 | 0.3 | 0.3 | 0.3 |
| Optical connector, return loss | 20 | 20 | 35 | 35 |

Note: all cable measurements are dB/km.
All figures are the same for ISO 11801 2nd edition and TIA/EIA 568B except where identified by 'TIA' or 'ISO'.

Standard and referenced as FOCIS 3P-0-2-1-1-0 for single mode plugs, FOCIS 3P-0-2-1-4-0 for multimode plugs and FOCIS 3A-2-1-0 for adaptors.

Where duplex connectors are specified, they must be polarised so that A goes to B and B goes to A. This will be achieved principally by the use of a crossover patch cord as shown in Fig. 4.24.

Both Standards state that there should be a colour coding scheme on the 'visible' part of the connector bodies and adaptors:

- Multimode: beige.
- Single mode: blue.

In addition, ISO 11801 2nd edition adds that single mode APC (angled physical contact) connectors shall be green. Most optical connector ferrules meet head on but this can lead to an unaccept-

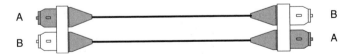

**Fig. 4.24** Duplex optical connector and patch cord.

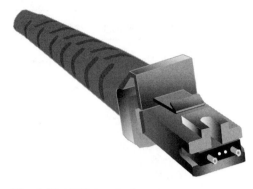

**Fig. 4.26** MT-RJ optical connector.

**Fig. 4.25** LC optical connector.

able level of light being reflected back into the transmitter (return loss) under some circumstances. Polishing the end face of the connector to about 8° from the perpendicular redirects most of the reflected light harmlessly into the cladding. This has been an issue for connectors close to high powered lasers, especially where analogue video transmission is involved. It is very rare, however, for this to be required in conventional LAN and premises cabling installations.

Some of the new connectors in circulation are the MT-RJ, the LC, the Fiber Jack and the SG or VF-45 from 3M. These connectors offer at least twice the packing density of the more established connectors like the SC and ST. Also more recent are the MU and the LX-5 connectors. Figure 4.25 shows an LC connector and Fig. 4.26 shows an MT-RJ optical connector.

It must also be recognised that other standards bodies have their own opinions about optical connectors, especially LAN standards such as IEEE 802.3z (gigabit Ethernet) and the ATM Forum.

**Termination methods**   Choosing an optical connector is one thing but how it is fitted to the optical fibre is another issue and there are seven different ways of doing it. The method of termination will have a big impact upon the quality of the termination, time spent on-site and, of course, overall cost. Usually more expensive connection methods mean less time spent on-site and vice versa.

- *Heat-cured epoxy*. This is sometimes called 'pot-and-polish'. The fibre is first stripped of its primary coating down to the 125-μm cladding layer. Adhesive is injected into the back of the ferrule with a special hypodermic needle. Second, the fibre is pushed into the back of the ferrule until it protrudes from the front face. The connector is then placed in a specially constructed oven for about 10 minutes. The adhesive cures and sets permanently after this time. The protruding fibre end is cleaved off with a scribe and then the connector end face is polished on a series of abrasive papers. The different grades of abrasive paper have a smaller and smaller grit size until the end of the fibre is perfectly smooth and polished and level with the face of the connector. The fibre end is then inspected with a special microscope, such as a Priorscope®, and if it looks good optically it is passed as fit for purpose. The connector may also be optically tested by mating it with a known test connector and measuring the attenuation across the two end faces with a power meter and light source. The whole process, excluding the power meter test, takes about 12–15 minutes per connector. The end result, however, is a permanent and usually high quality optical termination with good thermal stability.
- *Cold-cure or anaerobic*. The fibre is stripped and an adhesive is injected into the back of the ferrule. But this time the adhesive is part of a two-component system. The adhesive is only cured when it comes into contact with an activator. After preparing the fibre and injecting the adhesive into the ferrule, the bare fibre is dipped into the liquid activator. Before the activator can evaporate, the fibre is pushed into the back of the ferrule. Within 10 s or so the activator and adhesive have reacted and set. The protruding fibre end is then cleaved and polished as before. There is no need for oven curing.
- *UV cured*. The fibre and adhesive are prepared as before but UV light cures the adhesive. Once the fibre has been pushed into the back of the ferrule it must be exposed to strong UV light. After a

few tens of seconds the adhesive is cured. The protruding fibre end is then cleaved and polished as before.

- *Hot-melt.* In this method the ferrule is preloaded with an adhesive which softens when heated and then hardens again at room temperature. The ferrule is first heated in a special oven until the adhesive becomes soft enough for the fibre to be pushed through. The connector is left until it has cooled down and the adhesive has set again. The protruding fibre end is then cleaved and polished as before.

- *Crimp and cleave.* A very different method, but one promoted for its speed, is called crimp and cleave. The prepared fibre is pushed into the back of the fibre and simply crimped there. The fibre is then cleaved and may be polished to a small extent. Some questions have been raised about the effect of long-term temperature cycling performance on this method. The coefficients of thermal expansion for glass and ceramic (or polymer) are different. As the temperature goes up and down an effect called pistoning may take place if the fibre is not permanently secured in the ferrule, i.e. the fibre will try and retract down the ferrule when it gets colder and will try and protrude beyond the ferrule when it gets hotter.

- *Splicing pre-made tails.* All the above methods are suitable for on-site termination by the installer. A different approach is to terminate the fibre, using any of the above methods, in a clean factory environment, onto a half a metre or more of tight buffered fibre. This is called a tail, a tail cable or a pigtail. The installer will then splice the tail, by fusion or mechanical means, onto the end of the main fibre. This method is preferred when the installation site is very dirty or otherwise difficult to work in and is the main method chosen for single mode fibre systems.

- *Hybrid mechanical splice connectors.* A mechanical splice is a precision made tube, 126–128 µm internal diameter, that allows two accurately cleaved fibres to be brought into contact with each other. The hybrid connector is a factory terminated connector with the fibre tail premounted into one half of a mechanical splice. On-site the installer merely has to cleave the optical fibre and push it into the other side of the splice where it is retained either by glue or by crimping. This method gives quick high quality results, but at a price.

## Optical patch cords and pigtails

A patch cord is a piece of optical cable with an optical connector at each end used to connect active equipment into the cabling system or to patch between two optical patch panels. It would be rare to see an optical patch cord more than 30m long as this would then be in the realm of field deployable cables. After such lengths, special cable and connector designs are required to cope with the stresses imposed upon the fibres where they go into the back of the optical connectors.

An optical patch cord could have different connectors on each end and this is likely to be more prevalent as new connectors come into the market and different hardware manufacturers make different selections for their own range of equipment.

An optical patch cord can be simplex, meaning one fibre connectorised at each end, or duplex, meaning two fibres. If the patch cord is duplex then the standards require it to be of the crossover type, that is A goes to B and B goes to A as shown in Fig. 4.24.

A pigtail is a length of fibre connectorised at one end only, usually under factory conditions. The pigtail can then be spliced onto the main cable on-site using mechanical or fusion splicing techniques.

A mechanical splice is a precision made tube that allows two accurately cleaved optical fibres to touch each other and transfer light from core to core with a minimal optical loss.

Once stripped, cleaved and cleaned, the fibre is inserted into the mechanical splice tube. The other fibre to be spliced is inserted into the other end until it butts against the first fibre. The accuracy and tolerance of the tube and the fibre will determine how well aligned the cores are and hence the resulting optical loss. To improve matters it is common to preload the tube with an index-matching gel. This is a viscous fluid with a refractive index much closer to that of glass than air. It prevents back reflections from the inevitable tiny air gap that will exist between the two fibres and so improves optical loss. The longevity of the index-matching gel is not accurately known and will be influenced by temperature, temperature cycling and humidity; it will certainly be many years but it cannot match the permanence of a fusion splice.

Finally, the mechanical splice will be provided with further mechanical means to hold the fibre permanently in place, either by means of clips, adhesive or crimping.

Fusion splicing works by bringing two prepared optical fibre ends together between two electrodes. An electrical arc is struck between the electrodes and the heat of the arc melts the two ends of the fibres. The fibres are gently pushed together at this stage so they fuse together to form a permanent, low-loss splice.

The fibre ends are prepared in exactly the same manner as for the mechanical splice, that is, stripped, cleaved and cleaned. The fusion splicer is fitted with precision 'V' grooves that hold the fibres while their end faces are brought together between the electrodes.

For multimode fibre both mechanical and fusion splicing will give attenuation in the order of 0.2 dB. ISO 11801 and TIA/EIA 568B allow 0.3 dB for every splice however it is made. Fusion splicing is usually superior for single mode fibre because of its active alignment systems.

Some users are concerned about the long-term degradation of the index-matching gel within mechanical splices, especially in areas of extreme temperature change and low humidity.

The main argument revolves around costs however. A mechanical splice will cost between £5 and £10 and every subsequent splice will also cost the same. There is also an initial layout on the fibre preparation and cleaving kit but this is mostly common to both methods. A fusion splicer will cost between £4000 and £14000 but from then on each splice will only cost pence, that is, the cost of the splice protector. It can easily be seen, therefore, that for anybody intending to do more than a few hundred splices a year it will be cheaper to invest in a fusion splicer.

For emergency repairs and remote station repair kits it is easy to make an economic case for mechanical splices, as is also the case for cable installers who may only be doing a few dozen splices a year.

Latest developments include mass splicing or ribbon splicing. 4-, 8- or 12-fibre ribbons can be spliced together in one go. This offers huge potential savings in labour costs in fibre-rich installation scenarios. Both fusion and mechanical ribbon splices are available.

## Optical patch panels and joints

**Optical patch panels** As previously noted, the ends of the optical fibres have to be terminated or connectorised to be usefully employed, and those terminations have to be organised and protected so that they present a reliable, useable and manageable presentation to the end user. In data communications and LAN cabling the optical cable nearly always ends in a patch panel, which can be rack or wall mounted.

Sometimes optical cables have to be joined together, if runs are very long, for example. This is commonplace in telecommunications where cable runs are always long and joints need to be made around every 2 km or so. In premises cabling, by definition, cable runs are much shorter and so optical joints are not as common.

An optical rack-mounted patch panel will usually conform to the standard 19 inch (483 mm) wide rack-fitting system. The height is measured in 'U', where 1U is 44 mm, and the depth is variable but needs to be 300–600 mm. Twenty-four optical fibres can usually be terminated in a 1U panel, rising to 48 if using SFF connectors. This will vary according to the kind of connector used and the point at which the density becomes so great that it becomes very difficult to get one's fingers around the connectors. Although rack mounting is most common there is no reason why a patch panel cannot be wall mounted.

There are two ways to terminate the optical fibre: either direct connectorisation, using one of the methods described earlier, or by splicing on a ready made tail cable. The splice and the spare fibre that goes with it need to be organised and protected. The splicing or direct connectorisation process involves a technician working in front of the panel to fit the connector or splice. He/she will need at least a metre of fibre to work with so that the work area can be set up at a realistic distance from the panel. A splice will therefore involve 2 m of fibre as well as the splice itself that needs to be organised and protected. Even with direct connectorisation there will still be a metre of spare fibre to consider. With either method the end result is a piece of fibre with an optical connector on the end. This connector is fitted to the rear side of the bulkhead adapter that is mounted on the front face of the patch panel. The end user will then be able to make a continuous optical connection through the fibre network when plugging an optical patch cord into the font of the panel.

A 24-fibre patch panel may have to accommodate 24 fusion/ mechanical splices and up to 48 m of bare fibre. A good patch panel should have the following features:

- *Cable support/glanding*. As the cable enters the back of the panel there must be a method of securing it so that the fibres, connectors or splices are never subjected to any mechanical load.
- *Splice protection*. Splices should be secured and protected by some device like a cassette system or even a simple row of clips.
- *Spare fibre organiser*. The spare fibre, up to 48 m, must be organised so that the minimum bend radius of the fibre is never infringed and changes and repairs can be made in the future. The fibre cannot just be left to find its own path within the panel. Traditionally, 62.5/125 fibre working at 850 nm has been very forgiving about conditions such as macrobending. Users moving onto single mode fibre and wavelength division multiplexing in the future will not be so fortunate. The most popular method of controlling the spare fibre is a pair of plastic crosses that allow the excess fibre to be wound round them.
- *Sliders*. Ideally the rack-mounted patch panel should be fitted with sliders so that it can slide out and be self-supporting. This makes the installation job much easier with more reliable results.
- *Eye safety*. Every patch panel should be marked with the appropriate laser warning symbols denoting the possible class of laser in use. With higher powered lasers it may be necessary to introduce a method whereby it is impossible or at least difficult for anybody to be in direct line of sight of an optical adapter. If there is no patch cord connected to the adapter then no-one can know if there is any invisible infrared laser radiation being emitted from the adapter. This may be a worse problem for premises cabling

than telecommunications because cable runs are so much shorter and there will be much more light power reaching the far end. Possible solutions include always capping the optical adapters when not in use or angling the adapters downwards to make direct line-of-sight very difficult.

Large telecommunications networks may have thousands of optical fibres arriving at one location, all single mode and working at 1310 and 1550 nm. The organisation of the splices, attenuators, splitters, optical amplifiers, wavelength division multplexers and connectors is critically important or the network will not be reliable. The passive termination hardware is also a very small percentage of the total and so there is relatively little cost pressure on this item. It seems unfortunate in premises cabling that although people may be spending a million dollars/pounds/Euros on the cabling infrastructure there is a great resentment about procuring patching equipment concomitant with the sophisticated tasks expected of the cabling system.

**Optical joints**   Optical cables are typically made in 2–4-km lengths. If the cabling run is longer than that, two cables will have to be joined or spliced together. Joints may also be required if large cables have to go round such tight bends that the requisite pulling force would be unacceptable. Broken cables will also need joints inserted in them to repair the cable. There are two kinds of optical joints, in-line and distribution.

An in-line joint simply joins two optical cables together. Joints usually have to go outdoors and reside within inspection chambers buried in the ground or else are mounted on telegraph poles. Rugged mechanical construction and water resistance are therefore essential design elements for all classes of cable joints. The joint has three main construction elements:

- *Chassis*. A rigid metal chassis is required to secure the cable strength members and to hold the splicing protection mechanism.
- *Splicing protection mechanism*. The splices and spare fibre must be protected and organised. This is usually done with a cassette-based system.
- *Closure*. A waterproof plastic enclosure must cover and protect

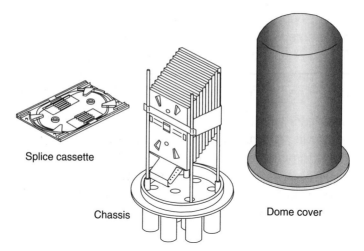

Splice cassette

Chassis                Dome cover

**Fig. 4.27** Optical joint.

the splices totally. The closure may be of heat-shrink material or two halves of a shell that bolt together.

The two cables may enter the closure from opposite ends but it is more common to have a dome-ended joint where both cables enter at one end and an elongated dome closure is placed over the chassis and sealed. Figure 4.27 shows a typical in-line optical cable joint.

A distribution joint, sometimes called a flexibility point, allows a high fibre count cable to be broken down into smaller cables. This is very useful in networks such as CATV distribution systems or the very large campus cabling systems that may be encountered on big military bases, universities and airports, etc. A small campus network using, for example, eight fibres would normally be configured as 'daisy-chaining' from one building to another. That is, the cable is terminated at each building and then patched onto the next leg of the cable leaving the building. This method would be totally uneconomic for an 80-fibre backbone cable that had to pass a potentially large number of users, as might be seen in a city centre-wide area network. In such circumstances it would be better to have a distribution joint mounted out in the street and to tap off a smaller cable, such as four or eight fibre, and drop it off into the users' premises.

The distribution joint takes advantage of the construction of multiple loose tube optical cables. An 80-fibre cable may be constructed of ten tubes with eight fibres per tube. In a distribution joint, the main cable is looped into the bottom of a dome-style closure, which has a special oval slot for such a purpose and is coiled inside, but it is not cut. Instead the sheath is removed to expose the fibre tubes. One or more tubes may be cut into and the fibres spliced onto a smaller drop cable. The splices are protected by a cassette-style mechanism. Up to six such cables may be dropped off in this manner from the main cable at one point. The distribution joint may feature other innovations such as hinged cassettes so that those splices can be re-entered in the future without disturbing existing circuits.

# 5

# Ensuring enough bandwidth for today and tomorrow

## 5.1 Introduction

Bandwidth is a measure of the capacity of a communications channel to carry useful information. The bigger the bandwidth the more information carrying potential that channel has. The units of bandwidth are the hertz (Hz) or kilohertz (1000 Hz) (kHz), megahertz (1000 000 Hz) (MHz) or gigahertz (1000 000 000 Hz) (GHz).

## 5.2 Digital communications

Users of communications systems are not really interested in bandwidth; they are more concerned with getting useful work, or data rate, from the communications channel such as the bits and bytes their computers can use. The units of data rate are bits per second (b/s) or kb/s or Mb/s and so on. If we are considering 8-bit words, or bytes, then the units are kilobytes per second (kB/s).

Digital data means that the units of information are one of two states, usually referred to as 'ones' and 'zeros'. Analogue information such as the human voice, music or video may be digitally encoded so that it is represented by a stream of ones and zeros. Ultimately it will take an analogue-to-digital converter to turn the digital information back into something we can make sense of.

It must be understood that megabits per second and megahertz are not the same, but they are related. The more bandwidth we have, the easier it is to send data down that channel.

Consider the bandwidth as a large highway where the number of traffic lanes represents the bandwidth. The useful work we get out of the highway is the amount of cargo carried by trucks down it. How can we get more cargo down the highway? Well, we can add more lanes; this is the same as adding more bandwidth. But we can also make the trucks go faster and make them carry more cargo. This is analogous to the numerous and sophisticated coding techniques that squeeze more data from restricted bandwidth channels.

In structured cabling the word 'bandwidth' is often used to describe the frequency range from zero to the point where the ACR (attenuation to crosstalk ratio) goes to zero. This use of the expression 'bandwidth' is not mathematically correct but it does give an idea of the useable frequency range of the cabling system. The useable frequency range, under the above definition, of structured cabling is:

- Class A: 100 KHz
- Class B: 1 MHz
- Class C: 16 MHz
- Class D: 100 MHz
- Class E: 200 MHz*
- Class F: 600 MHz

*Note that although Class E/Category 6 performance is specified up to 250 MHz, the positive ACR requirement only goes up to 200 MHz.

Many techniques are employed to squeeze as much data as possible down the bandwidth-limited cable communications channels. 1000BASE-T, gigabit Ethernet, for example, uses many devices to send 1000 Mb/s down a 100 MHz Category 5 channel. At the start the actual line rate is more likely to be 1.25 Gb/s, owing to the overheads of addressing, framing and error correction. This data stream is then split into four 300-Mb/s data streams, one for each pair of the cable, which are in turn encoded by a pulse amplitude modulation

(PAM) technique which sends one of five different voltage levels down the cable rather than the conventional two. Sophisticated circuits manage to send simultaneous bidirectional transmission down each cable pair. The receiver circuitry has to unscramble all this and rebuild the original 1000 Mb/s data stream. Such techniques are only possible with the advent of cheap microprocessors that can economically manipulate such vast quantities of digital signal processing data.

Optical fibre, metre for metre, has a much higher bandwidth than copper cable. The poorest multimode optical fibre has at least 20 times more bandwidth than Category 5 copper cable. Single mode fibre has a bandwidth thousands of times greater. The unit of bandwidth for optical fibre is still megahertz but it is conventional to use 'MHz.km' which gives an easier representation of how much bandwidth is available per unit length. For example, if an optical fibre is rated at 500 MHz.km, it means that in a 2-km link, 250 MHz of bandwidth is available.

Even when there is enough bandwidth, there must also be more signal than electrical noise or interference on the cable or the receiver will not be able to distinguish the desired signal from the background noise. Noise is unwanted addition to the desired signal. Noise is the limiting factor in the amount of data that can be transmitted through any channel. Noise can come from anywhere; it is *any* unwanted signal. White noise is energy randomly spread across a wide frequency band. In audio terms we would hear a rushing hiss that seemed to contain all the notes from high to low. We hear white noise when we tune off a station in the VHF band because the radio is giving us an audible interpretation of the white radio noise it is picking up and also creating within its own circuits. In a jet aeroplane the sound of the slipstream rushing over the fuselage gives a white noise sound within the cabin. It is called white noise after white light, which in reality contains all the colours of the visible spectrum (as demonstrated when white light is split into a rainbow by a prism) but which our eyes see as white light.

Noise can also be random impulse noise generated externally. Sources can include lightning, mobile phones, radar, fluorescent lights, any electrical switching device or even sun spots. Every active electronic device through which the signal passes will add noise, even an amplifier. An analogue signal cannot be regenerated an infinite number of times. Every time the signal passes through the amplifier circuit, a small amount of noise is added, which can never be removed. It is like making a copy of a copy of a copy of a VHS video cassette. Every copy will be slightly poorer than the one before. Digital signals can be repeated almost infinitely as long as the process happens before too much noise has been added. In the digital repeater a decision circuit will decide whether the input signal is a 'one' or a 'zero' (which it can do if not too much noise has been added) and then reproduce a brand new identical copy with no accumulated noise. Digital systems also have the benefit of error correcting codes that can make up for errors to some extent, but at the expense of sacrificing some of the system bandwidth.

A communications circuit will work when the size of the desired signal is very large compared with the interfering noise. A problem arises when the desired signal starts to become lost amidst the noise. The ratio of the desired signal to the interfering signal is called the signal to noise ratio (SNR) and its unit is the decibel.

Noise in a cabling system has three sources:

- *External noise*. This can be from anything, lightning, mobile phones, etc.
- *From within the same cable sheath or bundle*. Pickup from adjacent pairs or cables is called crosstalk. Crosstalk within the cable is the largest source of noise within a cabling system.
- From within the twisted pair itself. The main cause is return loss which will cause reflections back down the pair owing to impedance mismatch.

It is generally considered that for a cabling system to transmit a signal successfully, the size of the signal must not have been attenuated below the level of the background noise caused by near end crosstalk (NEXT) between the pairs. This is where the concept of a high (and positive) value of ACR comes from. It literally means that the desired signal has not been swamped by crosstalk noise which, as already stated, is always the largest source of noise in a cabling system.

## 5.3 Analogue communications

An analogue signal is an electrical or optical signal that continually varies in sympathy with the natural phenomena it is representing. The electrical current flowing around a loudspeaker is an analogue of the original acoustic signal. The electromechanical device of the loudspeaker turns the electrical current back into an acoustic signal that we can recognise. An analogue signal can have any one of an infinite number of states in between two limits. This is unlike a digital signal where the original analogue signal has been replaced by a number of binary words that have the same value as the original at any given measurement point.

Structured cabling is expected to carry analogue as well as digital data. Examples of this are ordinary telephones that require a 3-kHz signal bandwidth to operate. Simple closed circuit television (CCTV) surveillance cameras require about 6 MHz of bandwidth to operate. More sophisticated video systems require an even higher bandwidth. Video signals contain a huge amount of information and as such require a large bandwidth to transmit that information.

There are many types of video signal:

- CCTV surveillance, about 6 MHz required.
- Broadcast quality baseband video, about 12 MHz required.
- Frequency division multiplexed video, four channels would require about 40 MHz.
- RGB, red, green, blue CAD/CAM and dealer desk systems. About 30 MHz required in total.
- Broadband video means a number of video channels modulated onto a RF (radio frequency) carrier, like the signal coming off-air or through cable television coax. The bandwidth required is up to about 700 MHz.
- Video may already be digitised in which case it may require anything from 6 Mb/s to 200 Mb/s depending upon the quality required.

## 5.4   The role of wireless

Wireless links or radio LANs are often promoted as the natural progression of structured cabling systems, but this is not really the case. Wireless LANs have the great advantage of mobility and are ideal for:

* Hot-desk salespeople or other visitors.
* Warehouse operatives.
* Doctors/nurses doing the rounds.
* National monuments where cabling would be intrusive or destructive.

However, their disadvantages are:

* They can never compete with cable in terms of *bandwidth*. Hundreds of Mb/s are available through cable. Radio LAN users will be lucky to achiever hundreds of Kb/s or a few Mb/s. This, of course, has to be shared with all the other users in the same 'cell'.
* *Cost.* Wireless LANs invariably cost more to set up and maintain than cable-based LANs.
* *Security.* It is difficult but not impossible to eavesdrop on the information carried by other people's wireless LANs.
* *Denial of service attacks.* Wireless LANs can be disrupted by malicious attempts to jam them.
* *Health.* The long-term effects of dowsing office workers in low-level microwave radiation are not understood.

The role of wireless LANs will probably be such that they will offer mobility at the edge of a fixed structured cabling system. Cabling is always required, as the transmitter/receiver devices themselves have to be connected by cable back to the central equipment room.

## 5.5   What Category of copper cable to pick?

The choices of (copper) cable in the horizontal are Category 5, 6 or 7. Category 3 cable may be used for backbone telephone cabling. This gives the following system frequency ranges:

- Class C/Category 3: 16 MHz
- Class D/Category 5: 100 MHz
- Class E/Category 6: 200 MHz
- Class F/Category 7: 600 MHz

Multipair Category 3 cable is still the most cost effective for back-bone telephone cabling.

## 5.5.1   Category 5

For the horizontal, that is the Fibre-to-the-Desk cabling, if the user is happy to stay with the LANs of today, Class D/Category 5 will be adequate. This will supply up to 1000 Mb/s of data to the desk or up to 100 MHz of analogue video signal.

## 5.5.2   Category 6

On average, in the world of telecommunications and LANs, something new is discovered about every two to three years. If the user wants the cabling to be in place so that it can cope with the LAN and video systems of the next decade then Category 6 should be put in place. Category 6 is also becoming much cheaper and closer to the price of good quality Category 5. The argument may even arise for putting in Category 6 because the price differences are so low anyway. This is similar to the Category 4 versus Category 5 debate that briefly existed in the mid-1990s.

Another argument for Category 6 is the old truism that 'the better quality the cable then the cheaper the transmission electronics can be'. This is simply because use of a high quality, high bandwidth cable prevents the electronics from continually struggling to overcome the shortcomings of the cable and hence can be simpler and cheaper. This is typified by the 1000BASE-TX standard (ANSI/TIA/EIA-854), which promotes the use of gigabit Ethernet transmission on Category 6 only. The availability of twice as much bandwidth means that the expensive simultaneous bidirectional electronics seen in 1000BASE-T (IEEE 802.3ab) can be taken out of the chip.

Category 6 will also supply at least the twice the bandwidth for video systems.

### 5.5.3   Category 7

The future of Category 7/Class F is uncertain. At present its most enthusiastic proponents are in Germany, but it has generated little enthusiasm elsewhere. Category 7 will supply at least 600 MHz of system bandwidth, but this may be reduced by the quality of the connector. The RJ-45 version of the Category 7 connector struggles to maintain such a huge bandwidth capability when it started out in life merely to support a 3-kHz telephone signal.

Category 7 will probably be at least twice the price of Category 5 and possibly nearer three times. It will definitely be more expensive than FTTD and this is its greatest problem. Most users will see optical fibre as the next logical step when Category 6 can no longer satisfy their bandwidth needs.

Category 7 may be saved by the role of multimedia/broadband signals over structured cabling, a service it is well placed to support. This service will come at a price though. The cabling is already expensive and it will take additionally the service of two high quality video baluns to change the characteristic impedance of 75-$\Omega$ based video equipment to interface with 100-$\Omega$ structured cabling.

Two connectors have been approved for use with Category 7. One is compatible with RJ-45 8-pin modular connectors. The other is based on a totally different concept and is not compatible with RJ-45 connectors. There will not be much LAN equipment made with Category 7 interfaces.

Category 8 has also been proposed that will give a 1.2 GHz system capability!

The American (US) standards have reflected no interest in Category 7. The subject has only appeared in ISO 11801 and EN 50173.

## 5.6   Which optical fibre and how many?

### 5.6.1   Which optical fibre?

Picking the right optical fibre is much more simple with the advent of the OM1, OM2, OM3 and OS1 definitions. The user is now simply

Table 5.1 Fibre type required according to speed and distance

| Max. Transmission speed (Mb/s) | Transmission distance | | |
|---|---|---|---|
| | 300 m | 500 m | 2000 m |
| 10 | OM1 | OM1 | OM1 |
| 100 | OM1 | OM1 | OM1 |
| 1 000 | OM1 | OM2 | OS1 |
| 10 000 | OM3 | OS1 | OS1 |

invited to consider what transmission speed is required over what distance. This is summarised in Table 5.1.

The user is of course at liberty to have a mixture of fibres to cover a multitude of roles across the network.

A much considered question is how many fibres are required in the backbone cabling, presuming that the classical model of copper cable to the desk and fibre in the backbone is used. The question still has to be split into two parts, however: how many fibres link each floor to the main computer room (i.e. the building backbone), and how many are used for the interbuilding links (i.e. the campus backbone).

A communications link nearly always requires at least two fibres, a transmit and a receive. It is possible to put full duplex communications onto one fibre but this is not generally economic in terms of saved optical fibre versus increased equipment complexity. Some links, such as a CCTV link, will only require transmission in one direction. Some LANs, such as FDDI, require four optical fibres as they operate in a dual redundant ring format. This means that the information flow can go in either direction and in the event of a cable breakage the network can self heal and still operate effectively.

The best selling optical cable in Europe contains eight optical fibres and the logic behind this (although it may be long forgotten) goes along the following lines:

*I need at least two fibres, but I may require four for some types of LAN, installing more fibre is expensive so I will give myself 100% extra capacity for future expansion, hence I arrive at eight fibres.*

The reader may be interested to know that optical cable installations in Europe reveal the following statistics:[1]

- Average number of fibre links per site is 201.
- Median, i.e. the number where there are as many above it as below, is 24.
- Mode, i.e. the most commonly occurring number is 8.

It can be concluded from this that on the 'average' site three eight-fibre cables will be encountered. The average number of 210 is slightly misleading as it mixes up some very large FTTD installations with general run-of-the-mill backbone links.

Sometimes a more scientific rationale is needed to justify how many fibres are required and, of course, which type. In this case the intended transmission speeds and system loading have to be considered, but as a statistical approach to system loading has to be taken it eventually turns from an engineering decision to a business management one. The network manager must decide how much to spend to deliver the quality of service and achieve the level of system delay and response time that may be required. Table 5.2 attempts to

Table 5.2 Network utilisation and delay

| Backbone speed (Mb/s) | Number of users at 10 Mb/s | Average % utilisation of the network per user | Typical network delay |
|---|---|---|---|
| 100 | 10 | 100 | Zero |
| 100 | 100 | 10 | Average delay will be small |
| 100 | 1000 | 1 | Major delays at peak traffic times |
| 1000 | 100 | 100 | Zero |
| 1000 | 1000 | 10 | Average delay will be small |

indicate the amount of network delay relative to the network utilisation and available bandwidth. A simple backbone aggregate capacity that is simply the number of users multiplied by the maximum speed at which they can transmit will give no delay (at least not caused by the LAN) but will be expensive. As it is rare for all users to transmit data all the time, a more statistical approach is usually adopted.

A typical network usage with 10 Mb/s switched Ethernet to-the-desk is illustrated:

- General office environment, small emails and Word files: <1%
- General office user with occasional heavy traffic, e.g. large emails with attachments, Powerpoint and Access files: <2%
- Heavy user, desktop publishing, constant database access: 20–50%
- Computer animation, video streaming: 50–70%
- TV-quality video or video conferencing: 70–100%

To guarantee instant access for every user to the network service they require, the backbone speed must be the aggregate of the maximum data rate of every user, for example, a 100 Mb/s backbone link could not support more than ten users connected to it working at 10 Mb/s. However, as already stated, it is currently rare for general office users to make such heavy demands on their LAN and a statistical approach may be made. The list above and Table 5.2 show that a general office worker is probably only using about 1% of the LAN capability over a seven-hour shift. In such circumstances 1000 users could be connected to a 100 Mb/s backbone system using 10 Mb/s Ethernet switched links. Probably 99% of all emails or simple Word documents downloaded from the server would make the transit in an acceptable time, but occasionally several hundred users may all access the system at the same time and a message could take 100 times longer to transit the system.

The IT manager must decide what probability of success the company wants and can afford. It may not matter if occasionally workers have to wait 5 minutes for a small file download, but what will happen if the business is a call centre? What is the extra cost involved when operators are doing nothing and in addition customers become dissatisfied after waiting several minutes before their details appear on the screen?

The list above shows that heavy users may use the network up to 50% of the time, in which case a 100 Mb/s backbone should not have to support more than 20 10 Mb/s users.

Television quality video requires about 6 Mb/s, so putting surveillance cameras or distributing video channels (e.g. financial broadcast channels) over the LAN will quickly soak up the bandwidth. The same can be said of video conferencing, if using full motion and colour, this will also soak up megabytes of bandwidth. Some users with very heavy demands such as video streaming, video rendering, desktop publishing, medical imaging and computerised animation will already be on 100 Mb/s links anyway.

We have assumed a backbone speed of 100 Mb/s but it is possible to use gigabit Ethernet (1000 Mb/s) or even ten-gigabit Ethernet backbones. An upgrade path should always be available to the end user by using higher speed switches while still using the original pair of fibres. However, gigabit technology is expensive and for short links it may well be cheaper to use more fibres running many 100 Mb/s links in parallel rather than trying to achieve everything on just one pair.

In summary, there is no exact formula stating how many optical fibres should be installed, but it is possible to generalise. For building backbones the minimum should be eight optical fibres. If it is assumed that an average Ethernet chassis switch can handle up to 128 users then this can be equated with one fibre per 64 users attached to it. Eight fibres should therefore cope with up to 512 users running ordinary office applications. The next standard size cable is 12 fibres (64 × 12 = 768 users), then 16 (1024 users), then 24 fibres (1536 users). These figures are, of course, approximate; higher bandwidth users will need more fibres. The user always has the option of using blown fibre. A user can install a seven-way multiduct which is potentially an 84-fibre cable. Users of this technology tend to install the multiduct, blow in eight fibres for immediate use, and have the six spare tubes (72 more fibres capacity) available for further easy expansion.

# Reference

1   Elliott B and Gilmore M, *Fiber Optic Cabling*, 2nd edition, Butterworth–Heinemann, Oxford, 2001.

# 6

# The screened versus unscreened debate

Electromagnetic compatibility (EMC) is viewed as an electromagnetic emissions requirement whereas electromagnetic immunity (EMI) is a measure of the device or system's capability to reject outside electromagnetic interference. The term EMC is often used to cover both radiated and received electrical noise.

The UK Electromagnetic Compatibility Regulations (Statutory Instrument 1992/2372) enacted the European EMC Directive in UK law and as from 1st January 1996 information technology equipment must meet the relevant European EMC Standard and carry a 'CE' (Conformité Européene) mark indicating compliance.

The CE mark shows that a product meets all required European specifications, not just EMC. All active electronic equipment must carry the CE mark to demonstrate compliance. The CE mark, however, cannot be put onto passive equipment such as cabling and associated hardware. What kind of signal the cabling carries and the length and disposition of that cabling can change from one day to the next. Any EMC test done on structured cabling, for example a typical cabling layout running 100BASE-T Ethernet, is of general interest but it only proves that particular combination of manufacturer's equipment on that particular layout of cabling.

The philosophy must be that a well-balanced and engineered cabling system must not degrade the EMC performance of any active equipment connected to it. 'CE' conformant LAN equipment must still present a completely conformant IT system when plugged into the cabling system. A cabling system cannot rectify a badly balanced electrical signal and will radiate under such conditions.

ISO 11801 2nd edition invokes the following standards from Comité-International Spécial des Perturbations Radioélectriques (CISPR) as being the most appropriate for information technology equipment connected via structured cabling:

- CISPR 22: *Limits and methods of measurement of radio disturbance characteristics of information technology equipment.*
- CISPR 24: *Information technology equipment – immunity characteristics – limits and methods of measurement.*

European standards invoke the following:

- EN 50081: *Electromagnetic compatibility – generic emission standard.*
- EN 50082: *Electromagnetic compatibility – generic immunity standard – part 1: residential, commercial and light industry.*
- EN 50082: *Electromagnetic compatibility – generic immunity standard – part 2: industrial environment.*

American (US) standards would refer to Federal Communications Commission (FCC) regulations.

There are other tests specified from time to time but they are usually component tests and the tests listed above should take precedence as they relate to complete systems. Other such tests include:

- EN 61000 (IEC 61000): *Electromagnetic compatibility, Environment.*
- IEC 60801: *Electrostatic discharge and electrical fast transient immunity.*

IEC, CISPR, FCC and CENELEC (EN) standards address the issue of residential and industrial environment. EMC emissions are tighter for the residential environment than the industrial because it is considered that equipment will be located closer together in the residential environment. Class A is the non-residential environment and

Class B is the more demanding residential requirement. Both the LAN equipment and cabling infrastructure should meet CISPR, CENELEC and FCC Class B requirements when CE/FCC compliant transmission equipment is being used.

The cabling system should be able to withstand an impinging electromagnetic field of up to 3V/m across a band of DC to 1000MHz without incurring an additional bit error rate beyond that expected for any particular LAN protocol across all its operating speeds. For example, an ATM 155Mb/s signal sent with zero errors should not arrive with any more than one in ten to the power of ten errors when subject to the above interfering external signal.

How much 'real life interference' does 3V/m represent? Equation [6.1] approximates the field strength from a radiating source when the distance and power output of the source are known.

$$\text{Field strength} \approx \frac{\sqrt{30 \times W}}{d} \, V/m \qquad\qquad [6.1]$$

where W is the output power of source in watts and $d$ is the distance from source in metres.

(Another approximation sometimes used is $E = \frac{7 \times \sqrt{W}}{d} V/m$. The nature of the antenna, e.g. half-wave dipole, will also make a difference.)[2]

We can see that a mobile telephone (about 2W) from a few metres, and an airport radar (600kW) one thousand metres away, are capable of generating around 4V/m field strength.

Tests done by 3P Laboratories in Denmark,[1] whereby screened and unscreened cables were subject to 3V/m fields showed that peaks up to 35mV were induced into unscreened cables (UTP) but only a few millivolts was induced into a screened cable. Is this significant? For robust protocols, such as 10BASE-T, then it probably is not. But for the new multilevel coding schemes, like the 5 level PAM of 1000BASE-T, it may be.

Roughly speaking, for a 2-volt peak-to-peak signal at the transmitter and using a five level code, the difference between any two adjacent groups of code, that is, voltage levels, is going to be about half a volt. Given the attenuation of the cabling channel, what length of cable will it take to reduce the half a volt, or 500mV, to the level where it is of a similar magnitude to the 30mV or so that appear to make it into the cable with a 3V/m field impinging on a UTP cabling system?

To reduce 500mV to 30mV requires 24dB of attenuation. This just happens to be the channel attenuation of a 100m Class D channel at 100MHz. To reinforce the idea that 30mV is of an important magnitude we can see that the ATM 155Mb/s standard, *AF-PHY-0015-00,0* paragraph 5.3.1, calls for no more than 20mV noise to appear on the cable and the 1000 BASE-T standard *IEEE 802.3ab*, calls for no more than 40mV (40.7.5.1).

What happens if the field increases to 10V/m? We can presume the induced voltage in the cable will increase to around 100mV. It takes 14dB of attenuation to reduce 500mV to 100mV. The value of 14dB, at 0.22dB/m at 100MHz for the cable attenuation, means that the signal voltage differential has been reduced to the noise level at only 34m.

The practical effect of occasional noise of this magnitude would not be significant. A burst of noise will be masked by the error correcting codes of the LAN protocol. If it happened frequently then the user would notice longer and longer response times as data blocks were repeatedly retransmitted. But eventually the network would cease to function.

How will the users know if they are subject to a large interfering electromagnetic field? The practical effects, as we have said, will be long response times, system crashes, and for mainframe systems a mysterious logging on and off of terminals. But the only way to prove the existence of interference is to employ an EMC specialist who will set up antennas within the workplace and record the ambient electromagnetic noise environment across a suitable spectrum. The test should ideally be run across a full, normal, seven-day working cycle so that any level of interference coincident with some obvious event, such as a machine starting, can be seen.

Thus some conclusions can be drawn from these hypotheses:

- Users running their system in an environment of 3V/m or less ambient field, with only occasional transgressions above that, should be able to use a well-engineered and installed unscreened cable system at high LAN speeds up to and beyond 1000 Mb/s.

• Users operating in an environment of sustained interfering field of 3–10 V/m and wishing to run high-speed gigabit LANs and with distances in excess of around 40 m should consider a screened cable system.

• Users operating in an EMC environment above 10 V/m should consider using optical fibre.

Other practical issues to be taken into account are to install the cabling as far away from sources of interference as possible, for example fluorescent lamps, power cables, lift motors, switching gear, mobile 'phone users and any other radio or television transmitting equipment. See EN 50174 (discussed in Chapter 8) for more details of minimum spacing of cables from sources of potential interference.

The above discussion does not really point to the use of FTP or S-FTP or any of the other screened combinations available. We can say though that the more metalwork placed around the cable, the better the EMC performance will be, but this will come at the expense of more capital outlay, more space taken up and longer installation times. Users considering the use of analogue applications such as video would best be advised to use an STP/PIMF design of cable where each individual pair is screened from the others.

Any screened cable must be effectively earthed and bonded for it to function correctly.

# References

1   Bech E, 'Proposed cabling set-up for electromagnetic characterisation of cabling and EMC measurements on LAN systems', *Communications Cabling EC97*, ed AL Harmer, Amsterdam, IOS Press, 1997.

2   'Interference levels in aircraft at radio frequencies used by portable telephones', Report No. 9/40:23-90-02, Civil Aviation Authority, England, May 2000.

# 7

# Fire performance of indoor cables

Although PVC is the still the most common sheathing for indoor data cables, there is a choice of fire retardant PVC and a range of low flammability, zero halogen materials now available. Such materials have many interchangeable acronyms, such as LSF (low smoke and fume), LSF0H (low smoke and fume, zero halogen) and LS0H (low smoke, zero halogen).

The halogens are a class of elements consisting of:

- Chlorine.
- Fluorine.
- Bromine.
- Iodine.
- Astatine.

They are added to many plastics to act as stabilisers and flame retardants. Their combustion products give rise to many halogenated acidic compounds such as hydrochloric acid. The acidic fumes have a toxic effect on people and a very damaging effect on electronic equipment such as printed circuit boards. Many users, especially in Europe, are trying to do away with halogenated cables within buildings. Non-halogenated materials will still burn and give off noxious fumes such as carbon monoxide, but will not give off halogenated, acidic gases. A number of IEC standards relating to fire performance issues are:

- IEC 60332-1: *Flammability test on a single burning wire.*
- IEC 60332-3-24-c: *Flammability test on a bunch of wires.*
- IEC 60754: *Halogen and acidic gas evolution from burning cables.*
- IEC 61034: *Smoke density of burning cables.*

In Europe, at the time of writing, none of the above, or indeed any other, standards are compulsory under European law or directives. However, this will probably change when the Construction Products Directive (CPD) is published. Even so, IEC 60332-1 is seen as the base level of fire performance for intra-building cables. Note that halogenated material such as PVC can still meet tough fire tests, but cannot, of course, meet the zero halogen tests.

In the USA there are some very strict flammability standards for indoor cabling. The highest test is for cables placed in the plenum area. This is any area within a building that carries environmental air, such as the return airflow from an air-conditioned area. The moving air would make any fire in these spaces very dangerous and easily spread. A plenum space might typically be between the ceiling tiles and the deck of the floor above. Next rated below the plenum rated cables are the riser grade cables and below that are the general purpose cables. There is a substitution rule that allows plenum rated cables to be used in the riser, but riser cables could not be used in the plenum zone. Plenum cable has two disadvantages:

- They are expensive, typically three to four times the cost of PVC jacketed cables.
- They are expensive because they are jacketed with PTFE (polytetrafluoroethylene)/FEP (fluoroethylene polymer) (Teflon®) types of materials. This means that the product contains a halogen, fluorine, so when they do eventually burn they will give off halogenated acidic gases.

The USA NEC requires cables to be marked according to their classification, as shown in Tables 7.1 and 7.2, and permits substitutions of higher rated cables into lower rated environments. (NEC Tables 770-50, 770-53, 800-50, 800-53) as shown in Table 7.3.

Table 7.1  American (USA) NEC copper cable marking scheme

| Cable marking | Type |
| --- | --- |
| MPP | Multipurpose plenum |
| CMP | Communications plenum |
| MPR | Multipurpose riser |
| CMR | Communications riser |
| MP, MPG | Multipurpose |
| CM, CMG | Communications |
| CMUC | Undercarpet |
| CMX | Communications, limited |

Table 7.2  American (USA) NEC optical cable marking scheme

| Cable title | Marking | Test method |
| --- | --- | --- |
| Conductive optical fiber cable | OFC | General purpose UL 1581 |
| Non-conductive optical fiber cable | OFN | General purpose UL 1581 |
| Conductive riser | OFCR | Riser UL 1666 |
| Non-conductive riser | OFNR | Riser UL 1666 |
| Conductive plenum | OFCP | Plenum UL 910 |
| Non-conductive plenum | OFNP | Plenum UL 910 |

Table 7.3  Permitted cable substitutions (USA–NEC Code)

| Cable type | Permitted substitution |
| --- | --- |
| MPP | None |
| CMP | MPP |
| MPR | MPP |
| CMR | MPP, CMP, MPR |
| MP, MPG | MPP, MPR |
| CM, CMG | MPP, CMP, MPR, CMR, MP, MPG |
| CMX | MPP, CMP, MPR, CMR, MP, MPG, CM, CMG |

The relevant US fire tests are:

- UL 910    *Plenum test.*
- UL 1666    *Riser test.*
- UL 1581    *General purpose.*

The planned introduction of the Construction Products Directive into Europe will make a number of fire performance standards for building products, including cables, mandatory within the European Union. This has so far led to the development of *Euroclasses* of performance.

In developing the CPD and the Euroclasses, CENELEC have taken some of the IEC tests and added some of their own for European requirements:

- EN 50265-2-1: IEC 60332-1 *flammability, single cable.*
- EN 50266-2-4: IEC 60332-3 *flammability of a bunch of cables.*
- EN 50368: IEC 61034 *smoke evolution.*
- EN 50267: IEC 60754 *acidity and conductivity.*
- EN 50289-4-11: *Flame propagation, heat release, time to ignition, flaming droplets.*

The CPD proposals for cables within the European Union are shown in Table 7.4.

Table 7.4 shows that a Euroclass A cable can offer no contribution to a fire. This could only be achieved by the use of mineral insulated cables to which no current data cable could conform. Euroclass F has no fire performance requirements and could thus only be a cable

**Table 7.4 Construction Products Directive proposals**

| Fire situation | Euroclass | Class of product |
|---|---|---|
| Fully developed fire | A | No contribution to a fire |
| in a room | B | Very limited contribution to a fire |
| Single burning item | C | Limited contribution to a fire |
| in a room | D | Acceptable contribution to a fire |
| Small fire effect | E | Acceptable reaction to a fire |
| | F | No requirement |

used on the outside of a building or possibly buried in concrete or completely enclosed by a non-flammable conduit. Euroclass B would be the cable most closely resembling the US plenum style. At the time of writing it has yet to be decided by the European Union in which part of the building it is compulsory to use which type of cable. However, the overall cable rating would have to be as high as the requirements of the most onerous area that any length of the cable passed through.

ISO 11801 2nd edition does not make any recommendations about the fire performance of cables and leaves it up to the user to identify any national regulations. These are clear in USA but not elsewhere. Outside the USA the user should adopt IEC 60332-1 as the absolute *minimum* fire safety standard for indoor use and within the European Union follow the directions of the Construction Product Directive when it is published.

# 8

# Pathways and spaces

## 8.1   Introduction

The 'spaces' are enclosed areas that house cables, equipment and terminating hardware. Spaces are an essential and integral part of the structured cabling system. They include the telecommunications room, the equipment room and the building entrance facility. Figure 8.1 shows a representation of the possible spaces within a premises cabling system.

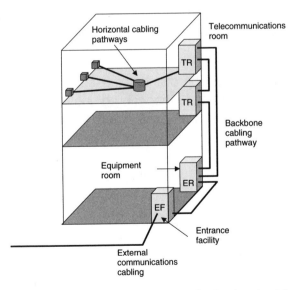

**Fig. 8.1** The 'spaces' within a generic structured cabling system.

Linking the spaces and also the TOs are the cable pathways. These are defined routes that the cables take. The pathways may be simply marked out routes or more substantial cable management hardware.

## 8.2 Definitions

The following definitions are based on expressions found in ISO 11801, EN 50174 and other standards.

### *Building entrance facility*

A facility that provides all necessary mechanical and electrical services to organise the entry of a cable into a building. This may also be a point of 'demarcation', that is, the point at which a clear transition occurs between the owner of one set of cable plant and another.

### *Distributor*

A collection of components such as patch panels and cords used to connect cables. There may be three types of distributor according to location:

- *Floor distributor* (FD). The distributor used to connect between the horizontal cabling and other cabling subsystems.
- *Building distributor.* A distributor where the building backbone cables terminate and connections to the campus cabling may be made.
- *Campus distributor.* The distributor from which all the campus cable emanate.

It should be noted that in the American TIA/EIA-568-B Standard, distributors are referred to as cross-connects, namely the horizontal cross-connect, building cross-connect and main cross-connect, respectively.

## Equipment room

A room dedicated to housing the distributors and application specific equipment such as router, servers, LAN switches, etc. It is presumed that only one equipment room would exist in a building although this may not always be the case.

## Telecommunications room

An enclosed space for housing telecommunications equipment, cable terminations and various distributors, interconnects and cross-connects. The telecommunications room has had several names during the evolution of the standards such as telecommunications closet and wiring closet. There would normally be at least one telecommunications room per floor, although small or sparsely populated floors may be served from a telecommunications room on another floor. In a small project the functions of the equipment room and telecommunications room may be combined in one room. ISO 11801 2nd edition states that for every 1000 square metres of office space there should be a telecommunications room.

## Cabinets, closures and frames

These are all devices for holding cable splices and termination equipment that can exist within the spaces and help to organise, present and protect the distributors:

- *Cabinets.* An enclosed equipment frame which may or may not have a door. The usable height of cabinets is usually given in units or 'U' (1 U = 44 mm) so that its equipment-carrying capacity can readily be determined. The mounting positions will normally conform to the 19-inch rack-mount Standard but other standards exist such as ETSI.
- *Frames.* An open construction of a cabinet. Frames can take up less floor space and give easier access to the equipment, but can look terribly disorganised unless cable management is strictly controlled.

- *Closure.* A fixture or fitting intended to house connecting hardware. The term is usually applied to cable joints and splices, but not exclusively.

## Zone

An area containing termination points that are served by a group of cabinets or frames.

# 8.3    What the standards say about the spaces

ISO and EN standards do not go into great depth about the requirements of the spaces. Lists of mostly common-sense requirements are given to the designer to consider when planning the spaces. The US standards go into more detail but the most detail can be found in the BiCSi design recommendations.[1]

Information on the spaces can be found in the following list of standards. A more complete list is given at the end of this chapter:

- ISO 11801 2nd edition.
- ISO 18010                     Information technology: Pathways and spaces for customer premises cabling.
- ISO/IEC TR 14763-2     Information technology – Implementation and operation of customer premises – Part 2: Planning and installation.
- ISO/IEC 15018            Information technology – Integrated cabling for residential and SOHO (small office, home office) environments.
- EN 50174-1                  Information technology – cabling installation – Part 1: Specification and quality assurance.
- ANSI/TIA/EIA-569-A     Commercial Building Standard for Telecommunications Pathways and Spaces.

Few hard numbers are given in these standards but one area that should be noted is the stated requirement for access room in front of cabinets and frames (or sides, if access is required), as the number given is slightly different for the different standards:

- ISO/IEC TR 14763-2   (5.2.2.2)   **900 mm** (it also states that no connection point should be above 2.5 m and none less than 0.15 m).
- EN 50174-1   (4.7.4 a)   **1200 mm**.

## 8.4   Essential design questions for the 'spaces'

Whereas some of the available literature, such as BiCSi manuals, attempts to prescribe all of the design and dimensional elements of the spaces, this is not the case in ISO and CENELEC standards. The few hard numbers which are given are stated above in Section 8.3. The design problem is left to the experience of the individuals concerned; this is appropriate when those individuals are themselves fully experienced in this aspect of design which covers structural design, electrical and communications engineering disciplines. We can assume, however, that this is rarely going to be the case.

To ensure that the formulated design is really going to address all the operational requirements of the spaces, it is best to start by posing all the necessary questions that will reveal these requirements. In this way the designer or architect can satisfy him or herself that the overall operational objectives will be met.

### 8.4.1   Location

Equipment and telecommunications rooms must be located to take the following factors into account:

- *Distance from the majority of the work area to be serviced.* The maximum horizontal cabling run allowed is 90 m. The telecommunications rooms must be specially located to ensure this distance limitation is met.
- *Relative vertical position of the telecommunications rooms.* Ideally the telecommunications rooms should be placed one above the

other in a multistorey building. This is to ensure that the backbone cabling is kept to a minimum and there are well-organised, accessible and spacious cable risers connecting them in the vertical plane.

- *Separation from sources of interference.* The rooms must be located well away from potential sources of electromagnetic interference such as lift shaft motors, electrical generating gear, television and cellular telephone transmitters, etc.
- *Accessibility.* The rooms must be located where it is possible to reach them with bulky and heavy equipment, e.g. a goods lift must be available with no intervening stairs.
- *The location must be environmentally secure.* For example, the rooms should not be below a water table if at all possible and must be impervious to all kinds of weather conditions.
- The building entrance facility must be located close to the actual point where the external cables enter and leave the building.
- The spaces must have direct access to the cabling backbone pathways with adequate space for all proposed cables and still maintaining all designated bend radii.
- Frames and cabinets must not be located in toilet facilities, kitchens, emergency escape ways, in ceiling or sub-floor spaces or within closures containing fire hoses or other fire-fighting equipment.
- The spaces must not contain any other pipework, plumbing or electrical equipment that is not directly related to the operation of the telecommunications or cabling equipment within that space.

## 8.4.2    Other design features

- *Security.* The rooms must be lockable and completely secure.
- *Accessibility to the rooms.* Doors large enough to wheel a 2-m equipment rack through must be available.
- *Power supply.* An adequate electrical supply must be supplied to the spaces and this must be presented with adequate mains power outlets for the equipment.

- *Uninterruptible power supplies, UPS.* It may be necessary to supply battery-backed UPS in the spaces.
- *Adequate floor space.* Sufficient floor space must be allowed for all of the equipment racks envisaged for the present and foreseeable future. This must include an allowance for the front access (at least 900 mm), other working areas, space for other equipment and UPS.
- *Adequate height.* The spaces must have sufficient height for the equipment racks to allow ventilation space and cable access above them. This means at least 2.6 m.
- *Adequate lighting.* There must be sufficient lighting installed that in all working areas it is possible to read black nine-point writing on a white background with ease.
- *Adequate floor strength.* The floor strength must be sufficient to cope with all of the envisaged equipment, including heavy UPS equipment with batteries.
- *Adequate ventilation and air conditioning.* If any active equipment is included in the spaces then they must be air conditioned with temperature of 18–24°C easily maintained under all load and external environmental conditions. Battery storage systems must also be adequately ventilated.
- *Wall boarding.* US standards insist on a thick plywood sheet applied to at least one wall to mount all the cross-connect equipment. This design feature should also be considered for many of the '110' and 'LSA' IDC terminations in use today.
- Building entrance facilities will usually require over-voltage protection equipment on copper cables entering the building.
- The earthing and equipotential bonding arrangements for all of the telecommunications equipment must be catered for.

### 8.4.3  Other control and monitoring features

The following points should also be considered by/with the end user:

- Access alarms on the doors.
- Motion/infrared detectors within the spaces.
- Fire and smoke detection within the spaces.

- Fire suppression system within the spaces.
- Flood detection within the spaces.

## 8.5    The pathways

The cable must have a defined route that will protect it according to the environment in which it exists. No standard exists that details exactly what kind of cable pathway is required in different circumstances and generally engineers are expected to use common sense to ensure the cable is adequately protected. What is clear, however, is that the cable routes cannot be abandoned to chance with installers allowed to pick any random route from A to B. There is also the commercial consideration that well-engineered cable management systems cost money and the end-users must decide how much cable management they want to pay for and the consequences of underinvestment in this area.

## 8.6    General rules concerning the installation of data cables

There are some general rules that apply to the installation of all cable bundles, regardless of containment type, and they are:

- *Cable ties must not be too tight.* Any cable within a tied bundle must be able to be moved through that tie with slight resistance. If the cable cannot move through the tie at all then it is too tight. Category 6 and optical cables cannot stand the same heavy duty 'lashing' as power cables. The tie must not be too thin either or it will start to cut into the sheath of the cable. Ideally the tie should be larger than 5-mm thick.
- *Cables must not be forced around bends tighter than that specified by the manufacturer.* Manufacturers will often specify six to eight times the cable diameter as the cable bend radius. This could range from 30 mm for 5-mm diameter Category 5 cable to 48 mm for 6-mm diameter Category 6 cable. ISO 11801 2nd

edition calls for 50-mm bend radius for four-pair cables and 15 times the diameter for cables with more pairs. This is a good generic specification to enforce upon cable contractors and should be easily achieved in main backbone and horizontal cabling runs. The last piece of cable jammed behind the TO however is always more problematic as there is never enough space available. At this point installers must ensure that they at least achieve the cable manufacturer's minimum bend radius requirements.

• *Cable bundles must not be too big.* There is no exact or correct value for the number of cables allowed in any one bundle but experience shows that between 24 and 48 cables is the optimum.

• Cables must not be dragged around or across sharp edges. This is self-evident.

• Power cables must cross data cables at right angles and should be separated from each other by a 'bridge'.

• Cables clipped directly to wall surfaces must be supported at 300 mm in unsupported horizontal runs, at 1000 mm for supported runs and 400 mm for vertical runs (BS 6701 7.2.5).

## 8.6.1   Separation of power and data cables

The separation of power cables from data cables is given in EN 50174-2: *Information technology – Cabling installation – Part 2: Installation and planning practices inside buildings*, and is summarised in Table 8.1.

From Table 8.1 we can see that the worst case, unscreened power cable next to unscreened IT cable is 200 mm. The standard makes no distinction between Category 5 or 6 and so on, or what kind of screening is used. It is either a screened or an unscreened IT cable. The table is mostly self-explanatory but some extra clarification is required:

• The metallic dividers are presumed to be earthed.

• The IT cables are balanced and working with a balanced transmission circuit.

• The power cables are standard 50–60 Hz operation with no unusual transients or high frequency or high voltage components.

Table 8.1 Cable separation rules from EN 50174-2:2000

| Type of installation | Separation distance | | |
|---|---|---|---|
| | Without a divider or with a non-metallic divider (mm) | Aluminium divider (mm) | Steel divider (mm) |
| Unscreened power cable and unscreened IT cable | 200 | 100 | 50 |
| Unscreened power cable and screened IT cable | 50 | 20 | 5 |
| Screened power cable and unscreened IT cable | 30 | 10 | 2 |
| Screened power cable and unscreened IT cable | 0 | 0 | 0 |

Note that EN 50174 permits no separation for the final 15m of the horizontal cable run.

- IT cables are expected to be at least 130mm from any kind of fluorescent lamps.
- Other separation distances may be required by other or national safety standards. Readers should consult national standards or IEC 61140 *Protection against electric shock* and IEC 60364 *Electrical installation of buildings*. Wherever there is a conflict between a safety standard and EN 50174 (which is mostly concerned about cable interference) then the higher figure for cable separation must prevail. British users must consult BS 6701.
- For backbone cabling the separation distances must be maintained end-to-end. For the horizontal cabling there is the practical problem of maintaining separations of up to 200mm in the shared trunking that is encountered around the walls of every office. To overcome this problem the standard adopts a more pragmatic approach.
- If the cable is screened and less than 35m in length, no separation is required.

- If the cable is longer than 35 m, the separation shall be maintained except for the last 15 m.
- The above two statements do not make it clear what happens to unscreened cable shorter than 35 m. The best interpretation of this clause (section 6.5.2 of EN 50174-2) is that for the last 15 m of the horizontal cable run, no separation is required between the data and power cables (apart from any other over-riding safety standard). Thus traditional office trunking can still be used. If we presume that we need around 3 m of cable to drop down the wall into the horizontal trunking system then a 12-m run is achievable. This actually equates to 24 m of wall if the cable run extends to 12 m in each direction away from the original cable drop.

TIA/EIA-569 and ANSI/NECA/BICSI 568-2001 also give separation distances between power and data cables. These are summarised in Table 8.2.

Table 8.2 ANSI/NECA/BICSI 568-2001 separation distances between power and data cables

| Condition | Minimum separation distance (mm) | | |
|---|---|---|---|
| | <2 kVA | 2–5 kVA | >5 kVA |
| Unshielded power lines or electrical equipment in proximity to open or non-metal pathways. | 127 | 305 | 610 |
| Unshielded power lines or electrical equipment in proximity to a grounded metal conduit pathway. | 64 | 152 | 305 |
| Power lines enclosed in a grounded metal conduit (or equivalent shielding) in proximity to a grounded metal conduit pathway. | — | 76 | 152 |
| Electrical motors and transformers | — | — | 1220 |

NECA = National Electrical Contractors Association.

Table 8.3 Worst case conditions of EN 50174 and BS 6701 overlaid on each other

| Type of installation | Separation distance (mm) | | | |
|---|---|---|---|---|
| | Without a divider | With a non-metallic divider | Aluminium divider | Steel divider |
| Unscreened power cable and unscreened IT cable | 200 | 200 | 100 | 50 |
| Unscreened power cable and screened IT cable | 50 | 50 | 50 | 50 |
| Screened power cable and unscreened IT cable | 50 | 30 | 50 | 50 |
| Screened power cable and unscreened IT cable | 50 | 0 | 50 | 50 |

Note that EN 50174 permits no separation for the final 15 metres of the horizontal cable run.

BS 6701 *Code of practice for installation of apparatus intended for connection to certain telecommunications systems*, also has requirement to separate power and communications cables. For power cable voltages not exceeding 600V AC, there must be a separation of at least 50-mm or else a non-conducting divider must be placed between them (7.7.4.1). For voltages above 600V AC the requirement is 150mm. This is a somewhat different approach to EN50174-2, which is concerned with interference between power and communications cables. BS6701 is more concerned with possible electrocution. Table 8.3 overlays EN50174 and BS 6701 to show all the requirements tabulated together.

BS 7671 *Requirements for Electrical Installations*, also known as the *IEE Wiring Regulations 16th Edition*, also defers to BS 6701 when considering the subject of the proximity of power cables to telecommunications cables (528-01-04).

# 8.7   Types of pathway – internal

### 8.7.1   Cable routes marked out on a floor

Apart from nothing at all, this is the simplest form of cable containment. It can be appropriate when cable can be laid directly on a relatively smooth concrete floor which in turn is covered by a false floor. The assumption here is that the cable is protected from damage by virtue of the floor suspended above it. Cables should be loosely bundled together with cable ties that can be nylon or Velcro-type or Millipede™ type.

### 8.7.2   Cable routes on the floor but with cable mat

Cable mat is spongy matting material that can be laid directly on rough concrete screed or wire basket tray to smooth out the mechanical load applied to points on the underside of the cables in contact with the floor or wire tray. Opinion is divided about whether this is really necessary or not. It would only be required on a concrete floor if the finish was extremely rough.

### 8.7.3   Cable conduit

Conduit is an enclosed tube that can be made of metal or plastic. It will provide maximum protection to the cable but will cost more to install and take longer to pull the cable into it. The cable should not take up more than 50% of the available cross-sectional area of the conduit. EN 50174 states that access to trunking, ducting or conduit systems should be available at intervals of no greater than 12 m. ISO 18010 (draft) *Information technology: Pathways and spaces for customer premises cabling*, gives a maximum figure of 30 m between pull points and not more than two 90° bends in the conduit. The inside bend radius of the conduit must not be less than six times the inside diameter of the conduit.

### 8.7.4　Tray, trunking, wireway and raceway

Exact definitions do not really exist, but their use implies a dedicated 'tray' that will hold the cables in place with easy access to those cables. Tray and trunking may be totally enclosed, in which case the term 'duct' or 'ducting' is more often used, especially when the containment is permanently or semi-permanently enclosed. The term 'cable tray' will be used here as the generic term for non-enclosed, rigid, cable support structures. The various forms and terminology are as follows with some forms shown in Fig. 8.2:

- Ladder: so called because the construction looks like a ladder. It is used mostly in vertical risers for cable support but can be used horizontally as well. Cable ladder implies a larger construction that may be used for many cables and/or larger communications and power cables.
- Solid-bottom cable tray.
- Perforated or trough cable tray.
- Spine cable tray (centre rail construction).
- Wire tray (welded wire construction).
- Mesh cable tray (wire or plastic mesh).
- Wireway (fully enclosed with gasketed cover).
- Cable runway (no side panels; cable is tie wrapped to base tray).

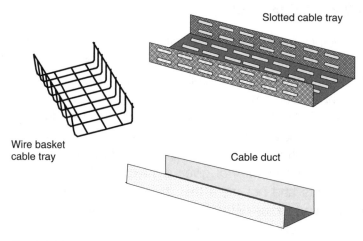

**Fig. 8.2** Forms of cable tray.

Which one is best really depends upon how much one wants to spend. The wire basket tray is suspected by some to cause problems with Category 6 cables and some manufacturers recommend cable mat to be placed in the bottom of the tray. No evidence to support this view ever seems to have been published. On the other hand manufacturers of cable tray have often published technical documents purporting to prove that wire tray is not a problem for high performance data cables.

For best EMC performance a solid metal trough will perform the best. The cables should not be piled higher than the side walls of the trough for any kind of cable tray. EN 50174 gives details of the earthing and bonding requirements of metal cable trough to ensure best EMC protection and in particular the requirements of joining two pieces of the trough together. A simple metal strap between the two may be fine for a safety earth but offers little in the way of EMC protection. For best EMC protection the shape of the trough must be maintained across each jointed section.

Power cables and IT cables should not be randomly mixed in the tray or trough but segregated with as much lateral separation as possible. Table 8.1 gives the cable separation according to EN 50174.

Cables should be bundled loosely together in the tray with cable ties in bundles of no more than about 48 cables. If the weight of the cable is supported in the horizontal then fixing to the tray needs to be at intervals of about 1–1.5 m. In the vertical plane the fixing should be approximately every 500 mm.

As with conduit, cable troughs should not be filled to more than about 50% of their theoretical maximum capacity. Remember that when calculating cable cross-sectional areas, the area taken is not $\pi r^2$ but the diameter of the cable squared. This is because each cable takes up the equivalent of a square of sides the same as the diameter of the cable, so a 6-mm diameter cable takes up 36 square millimetres of space.

To avoid crushing cables on the 'lower deck' of a large set of cable bundles each bundle should sit across the space made by the two bundles immediately below it, thus forming a pyramid effect. Figure 8.3 demonstrates this effect.

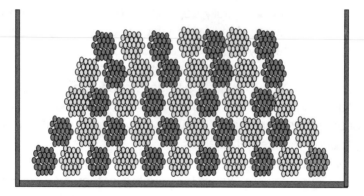

**Fig. 8.3** Cable bundles within a cable tray.

Alien crosstalk is sometimes mentioned in relation to cable bundles. Alien crosstalk is electromagnetic energy leaking from one cable and being picked up by another adjacent cable. In such circumstances it is the six cables in immediate contact with the victim cable (in the middle of the six) which will cause nearly all of the effect. Cables from further out in the bundle will have a very much smaller effect. For this reason the size of the bundle has little bearing on alien crosstalk problems. The main factor in increasing alien crosstalk is the presence of long parallel runs of the cables. These should be avoided if possible. In practice, individual four-pair cables do not run exactly parallel after they have gone around a few corners. However, bundled cables and preterminated cable bundles will tend to suffer the most from this phenomenon. If installers wish to use this method, they must obtain assurances from the cable manufacturer that their cables are suitable for bundling and pretermination.

With all cable trays care must be taken to ensure that right-angled bends in the tray do not end up forcing cables to go below their minimum allowed bend radius. Modern cable tray designed for data-communications and optical fibre cables now build in premanufactured corners.

EN 50174 requires an access space of 150-mm minimum above a cable tray to allow access.

## 8.7.5   Suspended cables

One method of containment is to suspend cables from hooks attached to the ceiling. These hooks are often referred to as 'J-hooks' as they look like a letter 'J'. J-hooks should have a broad base to support the cable and should support the cable at least every 500 mm. This method is popular in the USA but not often seen in Europe. If the size of the cable bundle is not too large then there is nothing wrong with this method except that BS 6701 appears to forbid it with the statement 'no ceiling hangers' in section 7.2.6.

## 8.7.6   Indoor cable duct

Underfloor duct systems are a cable management system embedded in the floor. Access or handhole units shall be placed in duct runs to permit changes in direction and provide access for pulling cables. The floor structure affects the type of underfloor duct system that can be accommodated in the floor and the total depth of concrete and method of pour will dictate the selection of the duct system. The various methods are:

- Monolithic pour.
- Slab-on-grade construction.
- Double-pour floor: the under-floor duct system is installed on the structural slab and the second pour buries the duct system.
- Prestressed concrete pour.

After the parallel under-floor duct runs have been established, the cross-runs of header duct and access units are determined by the density of the service requirements and the area to be supplied from each telecommunications closet. Provisions have to be made to connect the system to the telecommunications closets by a number of enclosed header duct home runs or a trench duct. Under-floor duct systems must terminate in the telecommunications closet with a slot or elbow as applicable.

Under-floor ducts mounted flush to the floor surface shall have removable cover plates through its entire length so that cables may be placed rather than pulled in. The cover plates shall have means for levelling to the intended finished floor surface and shall be designed to prevent entry of water into the duct.

The access or handhole unit provides access at the point of intersection of two or more under-floor ducts. In multiduct layouts, the interior of the access unit shall be partitioned to allow complete separation of the systems.

The cover plate shall be designed to prevent entry of water and shall have a means of levelling it to the surrounding floor area.

Service fittings, that is telecommunications, datacommunications and power, are available in several different types that serve from one to many different services. If electrical power is one of the services in a combined fitting, the fitting shall be fully partitioned.

## 8.7.7   Room perimeter pathways

Room perimeter pathways are usually surface-mounted trunking systems attached to the wall at desk level, sometimes called dado trunking, or at floor level, skirting trunking. These constructions are usually made from extruded white PVC and may be made up of several internal compartments. The internal compartments are for power and communications circuits. Some forms of trunking have metal separators between the compartments to improve EMC performance, and sometimes the plastic itself is metallised to achieve a similar, though not so effective, result. At appropriate distances, power and TOs are placed in the front of the trunking. These should be located within 3 m of where users are likely to want to make a connection.

A common mistake is to pick very slim trunking for aesthetic reasons. However, when it comes to installing all the power and data cables there is not enough room. One manifestation of this is crushed cables and reduced bend radii behind the outlet block, which usually extends deep into the trunking itself. Users must calculate the internal volume they require in the trunking after considering the density of outlets required, where they are going to make the drop to the rest of the horizontal cabling and the restrictions of EN 50174 concerning proximity of power and data cables.

Similar considerations must be given to the use of power poles that have exactly the same function as the perimeter pathway. All power poles must be securely mounted so that they cannot be knocked over.

### 8.7.8    Firestopping

There may be local or national regulations that require the maintenance of a firestopping system where cable trays, conduit or trunking and so on, pass through one fire zone to another. In such instances there will have to be some physical firstopping substance that stops flame or smoke or even the passage of air that may feed a fire further up the building.

This may be done by using several specialised materials now available and ordinary plaster or filler should not be used. There are numerous caulks, foams, putties, blankets, pillows, collars and blocks and so on, that are made just for the job and specialist manufacturers and contractors should be consulted. 'Intumescent' foam is often referred to. This is foam that will expand in the presence of heat to block off an air passage.

## 8.8    Types of pathway – external

There are four types of external cable pathways:

- Underground cable duct.
- Direct buried.
- Aerial.
- Fixed to the outside of buildings.

Some common aspects for all external cable routes are:

- Rights of way or easements. Do you have the rights to install cable in the planned location? To whom must you address the access permission?
- Shared route. If considering using an existing pathway, such as a duct, who owns it? Do you have permission to place your cables in there?

- Do the planned cable routes exist?
  - Who will plan them?
  - Is planning permission required?
  - Who owns the land where the cable pathway will go?
  - Who is going to fund the civil engineering works required?
  - Who is responsible for ensuring the cable pathway exists, is unblocked and available for use before the cable gang arrive on site?
  - Are draw pits and maintenance holes available or do they need to be built?

- Where are the other services, i.e. gas, electricity and water in relation to the cable pathways?
- Is the correct grade of cable being specified? Outdoor cable must be weatherproof and waterproof. Duct cable must be waterproof and strong enough to survive being pulled through up to 2 km of duct. Aerial cable must be ultraviolet resistant and capable of withstanding the great temperature swings it will encounter. Directly buried cable must be armoured and waterproof.

## 8.8.1   Underground cable ducts

Underground cable ducts are the most popular method of protecting external cables but some capital investment is required in order to have this asset available.

Cable ducts come in a range of standard sizes. They can be laid directly in a trench and backfilled with sand or soil or encased in concrete. Segments of duct should be joined together to keep out environmental influences but it must be accepted that in most countries the underground ducts will mostly fill up with water.

A cable duct is only useful if a method of placing the cable in the duct exists. This can be done in two ways, either by pulling the cable through the duct with a draw rope or by blowing it in with compressed air.

Basic guidelines should be followed for any type of cable installation, prior to cable placement:

- Undertake duct route survey.
- Determine the condition of the underground ducts.
- Develop a cable-pulling plan.
- Determine the pull cable lengths.
- Determine the locations of any intermediate pulling points.
- Determine the splice locations.
- Estimate the maximum pulling tensions and bend radii for each cable pull.

The installation team should be comprised of personnel familiar with all the aspects of a cable installation:

- Rodding and cleaning ducts.
- Cable pulling.
- Cable splicing.
- Splicing closures.
- Cable testing.
- Any special installation requirement.

External duct cables manufactured for duct installations are designed to withstand high mechanical stresses. In spite of the rugged construction, care must be taken to avoid cable damage during installation.

Precautions should be taken to prevent cable damage caused by any of the following:

- Exceeding the minimum bend radius.
- Exceeding the maximum pulling tension.
- Crushing the cable.
- Tearing the jacket over sharp objects.

## Bend radius

The cable minimum allowed bend radius at installation (under pulling load) and static (under no load) is specified in the respective manu-facturers' catalogues, installation manuals and cable drawings. As a general rule, the minimum bend radius allowed is:

Installation (under load):   16 × cable diameter.
Long term (no load):        10 × cable diameter.

## *Pulling tension*

The cable maximum pulling tension is specified in the respective manufacturers' catalogues, installation manuals and cable drawings. Consult these documents prior to installation in order to determine maximum pulling tension for the particular cable being installed.

## *Lubrication*

A cable lubricant is highly recommended to reduce friction at installation. The lubricant is to be applied as the cable enters the duct and at any pull-through manholes.

## *Crushing forces*

In order to prevent crushing the cable, enclose the work area where the cable is to be laid with barricades. When the cable has to be laid on the ground in high traffic areas, such as building entrances and road crossings, provision should be made for additional protection for the cable.

## *Cabling unoccupied duct*

Several methods may be used to rope an unoccupied duct prior to cabling operations:

- Magnesium duct rods.
- Fibreglass duct rods.
- Duct motor.
- Pneumatic roping equipment.

This operation should be performed well in advance of the cable pull so that removal of any obstruction encountered will not delay the cable installation. Once it has been established that the duct is clear and ready for the cable to be installed, a continuous length of 6-mm (approx.) polypropylene rope or other appropriate material should be pulled into the duct with the last pass of the cable mandrel.

## Cabling occupied duct

External duct cables can be installed in an occupied duct as long as the duct is not congested. In general it is not good cabling practice to install external duct cables into ducts where the total cable volume will exceed 50% of the duct space after the new cable has been installed. The clearance between the existing cables and inside duct wall must be greater than 40mm in order to allow enough room for the new cable to be installed.

When installing a cable in an occupied duct, care must be taken to prevent the new cable from twisting around the existing cables. A roping method using solid rods or a duct motor is therefore recommended.

## Checking for obstructions

Before the start of each installation, the duct should be checked for obstructions. For unoccupied ducts, a 75-mm by 25-mm cable mandrel should be pulled through the duct with no difficulty and with no evident scoring of the mandrel. For occupied ducts, the diameter of the mandrel should be 1.5 times the diameter of the cable to be pulled.

## Sub-ducts

The use of sub-ducts may be beneficial in instances where the existing duct contains known obstructions or where other larger cables will be installed at a later date. Sub-duct is usually also installed when small cables are going to be blown in.

Sub-ducts can be installed using the same procedures and tools used for cable placing. Because of the hollow nature of sub-ducts, greater care should be taken to prevent crushing or kinking at installation. Detailed installation practices should be obtained from the sub-duct supplier. The internal surface of the sub-duct is often grooved in the same axis as the duct itself to lower friction.

Before installing the cable in the sub-duct, the sub-duct should be anchored to the inspection hole cabling furniture at each end of the pull, using woven cable grips with an offset pulling eye.

## *Poisonous gases*

Asphyxiating or explosive gases can build up in underground ducts and maintenance chambers and any unventilated space connected to them. Installers must test underground spaces and connected unventilated spaces for such gases and take appropriate action to safeguard employees and members of the public.

Cable ducts entering a structure, such as the building entrance facility, should be sealed to prevent gases, water and rodents from entering the structure.

Power cables and communications cables must never share the same duct.

## *Direct buried cable*

A directly buried cable is laid into a trench in the ground with no other form of protection or ducting. The depth of the trench is likely to be between 600 mm and 1000 mm. BS 6701 calls for a minimum depth of 350 mm or 450 mm if the land is cultivated. This method can be cheaper than using cable duct but it means that the cable must be more robust than a duct cable and new cables cannot be added unless the trench is dug up again.

Long distance routes across country where no existing cable ducts exist, or where extra cables are unlikely to needed would qualify for direct burial.

The cable used must be very rugged which usually means armouring. The armour is needed to stop insect and rodent attack and incursions into the sheath by sharp stones and even agricultural disturbance of the cable route. Most armour consists of either steel wires helically wound around the cable under the outer sheath or a corrugated steel tape. No form of armour is going to protect the cable from the predations of a mechanical digger, however. Some areas might need other forms of protection such as oil or chemical resistance.

The trench is dug by hand or mechanical digger but there is a special machine called a mole plough that will plough a furrow, drop the cable into it and reclose the furrow again. If the soil is especially stony, the trench may be lined with sand before the cable is added and the cable might then be totally encased in sand before the general backfill is returned to the trench. Some users add cable tiles over the cable as an extra line of defence against being accidentally dug up. Another line of defence is to bury a brightly coloured plastic tape over the cable with appropriate warnings written onto it to the effect that a communications cable lies beneath it.

One military technique is to dig a broad trench and 'snake' the cable along the bottom, that is, following a path from one side of the trench to the other. This puts quite a large amount of cable into the trench compared with the actual distance being traversed from A to B. The point of this is that if a bomb blast occurs to the side of the cable trench then a huge amount of heave will be applied to the cable and a cable laid in a straight line will snap. But if it had been snaked into the trench there will be sufficient slack to allow the cable to move with the heave of the soil around it and survive the lateral forces applied to it.

As with cable ducts, which also have to be added to a trench, great care must be taken to avoid other services and utilities already in the ground. Cable joints and inspection chambers must also be designed into the route.

## 8.8.2   Aerial cables

Another technique is to string the cable between telegraph poles. The cable may be self-supporting or it may have to be secured to some form of support wire. The support wire may sometimes be called a catenary or a messenger wire.

Self-supporting cables must be especially designed for the job. If a support wire is used then the aerial cable may be the same product as a duct cable. The aerial cable must be designed to accept more severe environmental factors, however. An underground duct, apart from the constant presence of water, is a relatively benign place where the temperature does not alter too much from an average of about 10°C. An aerial cable must take the full brunt of the weather which includes large temperature swings, the thermal and ultraviolet impact of direct exposure to the sun, wind and ice loading and even possible shotgun damage! There is one design of aerial cable that comes complete with a Kevlar 'flak-jacket'!

Aerial cable must be fixed to the support wire at appropriate intervals of not more than 1.5 m. The height of the poles must be sufficient to allow clearance for people and vehicles passing underneath and adequate clearance must be given to any high voltage power lines in the vicinity. BS 6701 calls for span lengths of not more than 70 m with a minimum height above the ground of 3 m or 5.5 m where traffic may cross underneath. Telecommunications cables must always be below power cable when they share the same pole with a vertical separation of at least 1 m. A 2-m distance must be maintained between the aerial cables and high voltage lamps.

There is a special range of cables designed to be wrapped around high voltage cables. However, this is a specialist product that needs to be installed by specialist contractors and must not be attempted by the untrained.

The exact method by which the cable is going to enter the building must also be given detailed thought before the project starts.

### 8.8.3   Blown cable

Blown cable is a special cable designed to be blown into pre-installed ducts. This is often done for optical cables because many businesses do not want large amounts of capital tied up in optical cable plant, possibly many years before any revenue can be earned from that cable. Blown cable therefore requires the installation of a small cable duct, which can be viewed as a sub-duct, placed inside a larger cable duct. One manufacturer (MicroBlo™) offers a 48-fibre cable, 6 mm in diameter that can be blown into over 1000 m of 10-mm sub-duct.

The machinery required to blow the cable into the duct consists of a machine to feed the cable into the duct and a compressor to blow air over the cable and help move it along the duct. One version of this system aims to put optical cable into existing sewer pipes. This is sometimes called a 'no-dig' solution because it provides the important rights-of-way across a city centre without having to dig anything up. In such environments not only is it extremely expensive to undertake such enterprises, but planning permission from the local authorities may never be forthcoming.

One no-dig solution uses a robot to drag a special liner through a sewer pipe. Inside the liner are embedded several of the sub-ducts. Hot water is pumped through the centre of the liner causing it to be pressed against the inside wall of the sewer pipe. The heat of the water melts a resin in the liner and the liner is permanently stuck to the inside of the pipe. Cable or fibres can then be blown down the ducts at leisure at some time in the future.

# 8.9  Standards and pathways

The following are standards that relate to the subject of cable pathways:

- ANSI/NECA/BICSI 568-*2001*: *Installing Commercial Building Telecommunications Cabling.*
- ANSI/TIA/EIA-569-A: *Commercial Building Standard for Telecommunications Pathways and Spaces.*
- ANSI/TIA/EIA-570-A: *Residential Telecommunications Cabling Standard.*
- BS 6701: *Code of practice for installation of apparatus intended for connection to certain telecommunications systems.*
- BS 7671: *Requirements for electrical installations*, also known as the *IEE Wiring Regulations 16th edition.*
- IEC 60364: *Electrical installation of buildings.*
- IEC 61140: *Protection against electric shock.*
- ISO 11801 2nd edition: *Information technology – cabling for customer premises.*
- ISO/IEC 15018 (draft): *Information technology – Integrated cabling for residential and SOHO (small office, home office) environments.*

- ISO 18010 (draft): *Information technology: Pathways and spaces for customer premises cabling.*
- ISO/IEC TR 14763-2: *Information technology – Implementation and operation of customer premises – Part 2: Planning and installation.*
- EN 50085: *Cable trunking systems and cable ducting systems for electrical installations.*
- EN 50086: *Conduit systems for electrical installations requirements.*
- EN 50174-1: *Information technology – cabling installation – Part 1: Specification and quality assurance.*
- EN 50174-2: *Information technology – cabling installation – Part 2: Installation and planning practices inside buildings.*
- EN 50174-2: *Information technology – cabling installation – Part 3: Installation and planning practices external to buildings.*
- EN 61537: *Cable tray and cable ladder systems for electrical installations.*

# References

1   *Telecommunications Cabling Installations,* BICSI, McGraw-Hill, 2001.

# 9

# Earthing, grounding and bonding

## 9.1 Introduction

Earthing, grounding and bonding essentially covers all aspects of ensuring that all exposed and other extraneous conductive surfaces are connected to earth and all cables screens are all effectively earthed.

All exposed conducting surfaces and other extraneous conductive parts must be connected to an earth point for safety purposes. This is to ensure that if by accident a live conductor touched these parts the resulting circuit made to earth would cause such a large fault current that a fuse or a circuit breaker would blow somewhere. If the extraneous conductive part was not connected to earth, or left 'floating', it would maintain the live voltage on it until somebody came along and touched it and then that person's body would form the conductive path to earth.

In a structured cabling system the equipment racks, the active equipment, the metal patch panels and conduit, tray and trunking would all be considered as extraneous conductive parts and must be effectively earthed.

In a screened cabling system all of the screening elements of the cables, patch panels and connectors must also be earthed for the screening process to be effective. 'Floating' cable screens would be completely ineffective against interference and would also be considered to be a hazard as extraneous conductive parts.

The point of having electrical bonding systems within buildings that use information technology equipment may be summarised as providing:

1  safety from electrical hazards.
2  reliable signal reference within the entire information technology installation.
3  satisfactory electromagnetic performance of the entire information technology installation.

## 9.2  Definitions

There is some difference between US and standard UK English terminology presented here, as can be seen in the title of this chapter, 'Earthing, grounding and bonding'. 'Grounding' is an American expression not often seen in European standards where the expression *earthing* is more common. There is also a difference in screened cable terminology. North Americans would tend to use the term 'shielded' (although the USA Standard TIA/EIA-568-B uses the abbreviation ScTP to describe them), whereas standard (UK) English users would call them 'screened cables'. Both expressions amount to the same thing.

### *Bonding*

*Bonding* is often taken to mean earthing, but it really means more than just connection to earth. *Earthing* might imply a relatively high impedance path to ground that is good enough for safety purposes but not good enough for effective EMC/EMI control in a telecommunications system. The better expression to use in this context is *equipotential bonding*. However, in two US definitions of bonding one makes no reference to the equipotential aspect whereas the other one does: 'The permanent joining of metallic parts to form an electrically conductive path that will ensure electrical continuity and the capacity to conduct safely any current likely to be imposed' (TIA/EIA-607) or bonding: 'The permanent joining of metallic parts to form an electrically conductive path that will assure electrical continuity, the capacity to safely conduct any current likely to be imposed, and the ability to limit differences in potentials between the joined parts' ANSI/NECA/BICSI 568-2001.

## Earthing

*Earthing* is connection of the exposed conductive parts of an installation to the main earthing terminal of that installation (BS 7671).

## Equipotential bonding

*Equipotential bonding* is electrical connection putting various exposed conductive parts and extraneous conductive parts at a substantially equal potential (EN 50174-2).

## Grounding

*Grounding* is a conducting connection, whether intentional or accidental, between an electrical circuit (e.g. telecommunications) or equipment and the earth, or to some conducting body that serves in place of earth (TIA/EIA-607).

Interestingly, the BICSI TDM manual, 9th edition, gives a 'European' grounding definition: 'The process of interconnecting conductive structures that provide a Faraday cage (electromagnetic shield) for electronic systems. The term "electromagnetic shield" denotes any structure used to divert, block, or impede the passage of electromagnetic energy.'

## 9.3   The requirements of ISO 11801 2nd edition

ISO 11801 2nd edition makes some comments about earthing and bonding and also states that it must be in accordance with IEC 60364 or applicable national codes. Clause 11 of ISO 11801 2nd edition deals with screening and earthing and the following paragraphs summarise this clause.

### 9.3.1   General

Only basic guidance is provided. The procedures necessary to provide adequate earthing for both electrical safety and electromagnetic performance are subject to national and local regulations, always to proper workmanship in accordance with ISO/IEC TR 14763-2, and in certain cases to installation specific engineering. Some cabling employs components that utilise screening for additional crosstalk performance and is therefore also subject to screening practices. Note that a proper handling of screens in accordance with ISO/IEC TR 14763-2 and suppliers' instructions will increase performance and safety.

### 9.3.2   Electromagnetic performance

Cabling screens should be properly bonded to earth for electrical safety and to optimise electromagnetic performance. All cabling components that form part of a screened channel should be screened and meet screening requirements. Cable screens shall be terminated to connector screens by low impedance terminations sufficient to maintain the screen continuity necessary to meet cabling screening requirements. Suppliers' instructions about how to make low impedance terminations shall be asked for and observed. Work area, equipment cords and the equipment attachment should be screened and shall provide screen continuity.

### 9.3.3   Earthing

Earthing and bonding shall be in accordance with applicable electrical codes or IEC 60364-1. All screens of the cables shall be bonded at each distributor. Normally, the screens are bonded to the equipment racks, which are, in turn, bonded to building earth. The bond shall be designed to ensure that the path to earth shall be permanent, continuous and of low impedance. It is recommended that each equipment rack is individually bonded, in order to assure the continuity of the earth path.

The cable screens provide a continuous earth path to all parts of a cabling system that are interconnected by it. This bonding ensures that voltages that are induced into cabling (by any disturbances) are directed to building earth and so do not cause interference to the transmitted signals. All earthing electrodes to different systems in the building shall be bonded together to reduce effects of differences in earth potential. The building earthing system should not exceed the earth potential difference limits of 1 volt rms (root mean square) between any two earths on the network. The root mean square is like an average because it can mean any waveform whose total energy content is the same as would be delivered by 1 V DC.

## 9.4   Applicable codes and standards

The basic job to be done is to supply a low impedance bond to earth in order to achieve not more than 1-V rms potential difference between any two points on the earth system.

ISO 11801 refers to *applicable electrical codes* as well as IEC 60364 in order to achieve this. Also, remember that there are two types of earth connection, the *protective earth* (PE), whose function is to prevent exposed and extraneous conductive surfaces from carrying fault voltages and currents, and *functional earths* (FEs).

The PE is formed by a protective conductor whose main job is to prevent an electric shock and is intended to connect together the following parts:

- Exposed conductive parts.
- Extraneous conductive parts.
- The main earthing terminal.
- The earth electrodes.
- The earthed point of a neutral conductor.

A FE is provided only to enable equipment to function properly. It is not intended to offer protection either to the user or to the equipment. Examples of the uses of a FE are:

- To provide a zero-volt reference point.
- To enable an electromagnetic screen to be effective.

- To provide a signalling path for some types of communications equipment.

One conductor may provide both the functions of protective earth and functional earth but the requirements for protective measures must take precedence. So whereas a large protective earth conductor may be used for 'functional', that is telecommunications purposes, a (usually) much smaller conductor used as a functional earth must not be the only source of protective earthing.

The standards used are:

- ANSI/TIA/EIA-607: *Commercial Building Grounding and Bonding Requirements for Telecommunications*.
- BS 6701: *Code of practice for installation of apparatus intended for connection to certain telecommunications systems*.
- BS 7671: *Requirements for electrical installations*, also known as the *IEE Wiring Regulations 16th edition* and its corresponding *Guidance Note No. 5 Protection Against Electric Shock*.
- IEC 60364-1: *Electrical installation of buildings – Part 1: Fundamental principles, assessment of general characteristics, definitions*.
- IEC 60364-4-41: *Electrical installation of buildings – Part 4-41: Protection for safety – Protection against electric shock*.
- IEC 60364-5-548: *Electrical installation of buildings – Part 5: Selection and erection of electrical equipment – Section 548: Earthing arrangements and equipotential bonding for information technology installations*.
- ISO 11801: 2nd edition *Information technology – cabling for customer premises*.
- EN50310: *Application of equipotential bonding and earthing in buildings with information technology equipment*.
- EN 50174-2: *Information technology – Cabling installation – Part 2: Installation and planning practices inside buildings*.

The geographic targets of these standards are:

- ANSI/TIA/EIA        United States of America.
- British Standards   United Kingdom.

- CENELEC (EN)    European Union.
- ISO/IEC    Worldwide.

## 9.5  The electrical power distribution system

For any building of any size the electrical power supply arrives as 'three-phase'. In Europe this means three different conductors each carrying about 415 V, alternating current at 50-Hz frequency (in the USA it can be 208 or 480 V, 60 Hz). Each conductor carries a voltage 120° out-of-phase with the other two conductors. These are the 'live' conductors. Every live conductor needs a return path to complete the circuit and this is known as the neutral. The advantage of three-phase is that the shared current of the three phases returning down one common neutral is not merely the arithmetic addition of the three currents, as it would be if it was direct current. Because the currents are 120° out-of-phase, the total combined current on the neutral is much less. This means that instead of supplying the power on six large conductors (three live and three neutral) or the equivalent of two very large conductors, three phase is delivered on three live wires and one relatively small neutral conductor. This is shown in Fig. 9.1.

The higher voltage, 415 V, is also more efficient, as the higher the voltage, the lower the current needs to be in order to deliver the same amount of power. The size of the conductor needed is determined

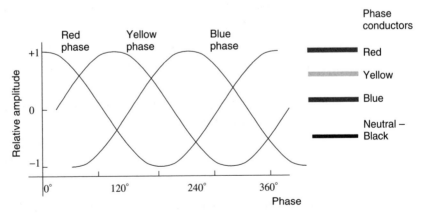

**Fig. 9.1** Three-phase supply.

by the size of the current flowing, so a higher voltage means a smaller conductor. This is why long distance electrical transmission uses extremely high voltages to reduce loss in the cable.

Although large electrical machines may take the three-phase supply directly, a transformer is required somewhere in the system so that the electrical supply delivered into the home or office is the more familiar 240 V.

The neutral conductor is earthed at some point, although exactly where depends on the exact design. Hence we see expressions such as TN-C and TN-S (see below) used to explain how the electrical distribution system is connected.

## 9.5.1   The electrical distribution codes

The codes used have the following meanings:

*First letter* – Relationship of the power system to earth:
- T = direct connection of one point to earth;
- I = all live parts isolated from earth, or one point connected to earth through an impedance.

*Second letter* – Relationship of the exposed conductive parts of the installation to earth:
- T = direct electrical connection of exposed conductive parts to earth, independent of the earthing of any point of the power system;
- N = direct electrical connection of the exposed conductive parts to the earthed point of the power system (in AC systems, the earthed point of the power system is normally the neutral point or, if a neutral point is not available, a phase conductor).

*Subsequent letter(s)* (if any) – Arrangement of neutral and protective conductors:
- S = protective function provided by a conductor separate from the neutral or from the earthed line (or in AC systems, earthed phase) conductor.
- C = neutral and protective functions combined on a single conductor (protective earth neutral) PEN conductor.

The common arrangements are TN, TN-C, TN-S, TN-C-S, TT and IT. IT in this instance means a system having no direct connection between live parts and earth, the exposed conductive parts of the electrical installation being earthed. **It has nothing to do with an information technology system.** A TN-C and a TN-S configuration are shown in Fig. 9.2(a) and (b) as an example.

A PEN conductor is not recommended for information technology installations as the PEN will carry a mixture of unbalanced currents and an accumulation of harmonic signals and other disturbances which may find their way into the IT equipment. So wherever possible the TN-S distribution method should be used:

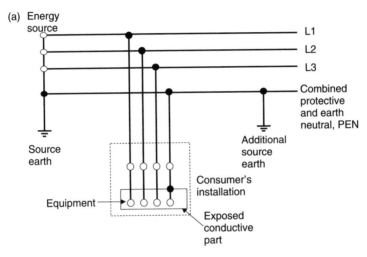

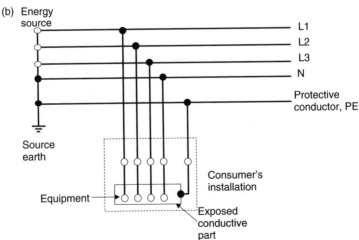

**Fig. 9.2** (a) TN-C electrical distribution system. (b) TN-S electrical distribution system.

## *Reference*

- EN 50174, section 6.4.3 (b).
- EN 50310, section 4.3.
- IEC 60364-5-548, section 548.4.
- BS 7671 (IEE 16th edition) nominates EN 50310 in matters referring to telecommunications.

The complete design of the electrical distribution system is beyond the scope of this book and more specialised help should be sought to address this area.

The separation of the power and communications cables is covered in Chapter 8.

# 9.6   The equipotential bonding system

Equipotential bonding is a series of electrical connections maintaining various exposed conductive parts and extraneous conductive parts at substantially the same potential. We can restate this by saying everything metallic, that is not part of a current carrying circuit, is connected to everything else and then to earth.

The idea is to form an earthed equipotential zone, in which, under fault conditions, the difference in potential between simultaneously accessible conductive parts will not cause electric shock.

In an installation, main equipotential bonding conductors shall connect to the main earthing terminal extraneous conductive parts of that installation including the following:

- Water service pipes.
- Gas installation pipes.
- Other service pipes and ducting.
- Central heating and air conditioning systems.
- Exposed metallic structural parts of the building.
- The lightning protective system.
- *The metallic sheath of any telecommunications cable.*

The size of the equipotential bonding conductor depends upon the size of the fault current it may have to pass.

For TN-S distribution systems:

Table 9.1 BS 7671 Cross-sectional area of main equipotential bonding conductors when using protective multiple earths

| Copper equivalent cross-sectional area of the supply neutral conductor (mm²) | Minimum copper equivalent cross-sectional area of the main equipotential bonding conductor (mm²) |
| --- | --- |
| 35 or less | 10 |
| 35 to 50 | 16 |
| 50 to 95 | 25 |
| 95 to 150 | 35 |
| Over 150 | 50 |

'the main equipotental bonding conductor shall have a cross-sectional area not less than half the cross-sectional area required for the earthing conductor of the installation and not less than $6\,mm^2$. The cross-sectional area need not exceed $25\,mm^2$ if the bonding conductor is of copper . . .' BS 7671

Table 9.1 (also from BS 7671) should apply where a protective multiple earth system is used as in TN-C-S.

The main equipotential bonding connection to any gas, water or other service shall be made as near as practicable to the point of entry of that service into the premises. Where there is an insulating section or insert at that point, or there is a meter, the connection shall be made to the consumer's hard metal pipework and before any branch pipework. Where practicable the connection shall be made within 600 mm of the meter outlet union or at the point of entry to the building if the meter is external.

## 9.6.1   Supplementary equipotential bonding

Supplementary equipotential bonding means extra connections between the extraneous and exposed conductive parts at a more local level. This is to reinforce the equipotential zone caused by the main protective conductors running back to the main earthing terminal. Exactly when and where supplementary equipotential bonding is to be used is difficult to define and relates to complicated rules about automatic disconnection times.

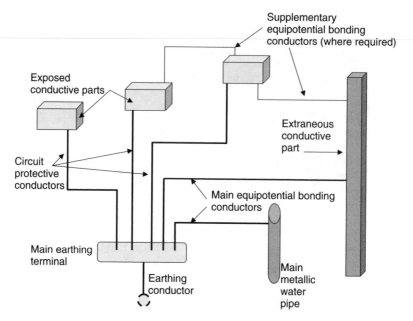

**Fig. 9.3** Typical arrangement of earthing conductors within a building.

The size of the supplementary equipotential bonding connector shall be not less than $4\,mm^2$ where no mechanical protection is provided or $2.5\,mm^2$ where protection is provided (BS 7671 547-03).

Figure 9.3 shows a typical arrangement of earthing conductors within a building.

### 9.6.2   Colour

The colour of all protective conductors shall be green and yellow (BS 7671 51A).

### 9.6.3   Large inductive ground loops

One standard in particular, EN 50174, notes the potential of electro-magnetic pickup when the area formed by a loop of protective earth conductors is relatively large. Supplementary equipotential bonding between equipment or racks should then be used to reduce the common impedance between the equipment.

## 9.7   Functional earthing and screened cabling

### 9.7.1   BS 7671 and IEC 60364

An FE is provided to enable equipment to function properly. A PE is there to provide protection for people, property and livestock. The FE is not intended to offer this kind of protection but to:

- Provide a zero volt reference point.
- Enable an electromagnetic screen to be effective.
- Provide a signalling path for some kinds of equipment.

Where earthing for combined protective and functional purposes is required, the requirements for protective measures shall take precedence.

In every installation a main earthing terminal shall be provided to connect the following to the earthing conductor (BS 7671 524-04):

- the functional earthing conductors,
- the circuit protective conductors,
- the main bonding conductors and
- the lightning protection system bonding conductor.

BS 7671 defers to EN 50310 for any further details about FEs.

ISO 11801 2nd edition refers to IEC 60364 for comments on earthing systems. *Part 5: Selection and erection of electrical equipment – Section 548: Earthing arrangements and equipotential bonding for information technology installations* is the relevant section of that particular Standard. The Standard offers the following definitions:

- *548.1.3.1 functional earthing* – The earthing of a point in a system or in an installation or in equipment, which is necessary for a purpose other than protection against electric shock [Future IEV 195-01-14].
- *548.1.3.2 functional earthing conductor* – An earthing conductor provided for functional earthing [Future IEV 195-02-14].
- *548.1.3.3 functional earthing and protective conductor* – A conductor combining the functions of both protective earthing conductor and functional earthing conductor [Future IEV 195-02-15].

- *548.1.3.4 earthing bus conductor* – A conductor (or busbar) connected to the main earthing terminal.

Functional earthing may be provided by using the protective conductor of the supply circuit for the information technology equipment. In some cases, the functions of the functional earthing and protective conductor are provided by installation of a separate dedicated conductor connected to the main earthing terminal of the building.

The following conductors are permitted to be connected to the earthing bus conductor:

- Conductive screens, sheaths or armouring of telecommunication cables or telecommunication equipment.
- Earthing conductors for overvoltage protective devices.
- Earthing conductors of radio communication antenna systems.
- The earthing conductor of an earthed DC power supply system for information technology equipment.
- Functional earthing conductors.
- Conductors of lightning protection systems (see IEC 61024-1).

It is recommended that for racks or rows of cabinets of 10 m or more in length, the functional earthing and protective conductors should be connected at both ends to the local equipotential bonding mesh or the earthing bus conductor. When a separate earth electrode is installed for functional earthing purposes it shall be connected to the main earthing terminal of the installation by a functional earthing conductor with a minimum cross-sectional area of 10-mm$^2$ copper or equivalent conductance.

IEC 60364-5-548 describes three methods for earthing and improving equipotential bonding to achieve electromagnetic compatibility.

## Method 1: Radially connected protective conductors

Method 1 uses the normal protective conductors associated with the supply conductors. The protective conductor at each equipment provides a relatively high impedance path for electromagnetic disturb-

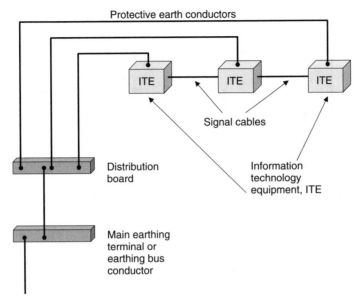

**Fig. 9.4** IEC 60364-5-548 Method 1 Radially connected protective conductors.

ances (other than mains-borne transients) such that interunit signal cables are subject to a large proportion of the incident noise. Equipment must therefore have a high immunity to function satisfactorily.

By providing a dedicated branch of the supply circuit and earthing system serving the information technology equipment, segregated from other supply circuits and earthing systems and extraneous metalwork, incident disturbances can be much reduced. In some cases the main earthing point of the radially connected functional earthing and protective conductors for the information technology equipment may be earthed by a separate dedicated insulated conductor connected to the main earthing terminal (see Fig. 9.4).

## Method 2: Local horizontal equipotential bonding mesh

The normal protective conductors are supplemented by equipotential bonding the components of the information technology system to a local mesh. Depending on the frequency and the mesh spacing this

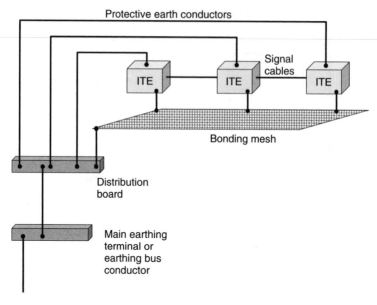

**Fig. 9.5** IEC 60364-5-548 Method 2 Local horizontal equipotential bonding mesh.

can provide a low impedance reference potential plane for signal interconnections between those system components in the close proximity to the mesh.

As with method 1, additional immunity may be provided by segregating the whole of the information technology supply circuits and earthing system, including the bonding mesh, from other supply circuits and earthing systems and extraneous conductive parts such as building metalwork (see Fig. 9.5).

## Method 3: Horizontal and vertical equipotential bonding mesh

In this case the normal protective conductor arrangements are enhanced by providing equipotential meshes on each floor. These in turn have multiple bonds to building metalwork, the exposed conductive parts of the electrical installation and metalwork of other services. Vertical equipotential bonding connections between floors may be provided. This method of earthing may also employ a ring

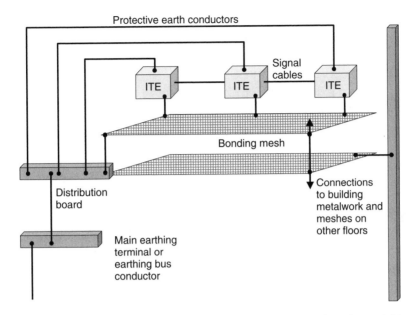

**Fig. 9.6** IEC 60364-5-548 Method 3 Horizontal and vertical equipotential bonding mesh.

earthing bus conductor extending the main earthing terminal of the building (see Fig. 9.6).

Method 1 is the most easily implemented, especially in existing buildings. The difficulty and cost of implementation increase through method 2 and method 3. However, these methods are more likely to provide an acceptable environment for unspecified future information technology equipment.

## 9.7.2  BS 6701

BS 6701: *Code of practice for installation of apparatus intended for connection to certain telecommunications systems* says more about functional earthing than most standards.

BS 6701 states that the FE should be connected to earth only at one of the following:

- The consumers' main earthing terminal (CMET).
- A remote dedicated FE terminal directly connected to the CMET

by a conductor of not less than 1.5 mm² or as specified by the equipment manufacturer, whichever is greater.

- An earth terminal at a power main area distribution panel in accordance with BS 7671.
- An external FE terminal connected to the earth terminal above.
- An earth electrode system buried in the ground as recommended in BS 7430. In this case a bonding conductor should be connect the earth electrode and the CMET.

The FE conductor should be an insulated copper wire with a minimum cross-sectional area of 1.5 mm². This conductor may be a separate cable or a suitable conductor in a multiconductor cable (for example British Telecom internal cables CW1308 contain an integral 1.38-mm diameter conductor, i.e. 1.5 mm²). Where the FE conductor is provided as a separate cable, the sheath colour should be cream. The sheath should be embossed with the words 'TELECOMMS FUNCTIONAL EARTH'.

Where the FE is terminated at the CMET or elsewhere a permanent label should be used with the words 'TELECOMMS EARTH DO NOT REMOVE'.

## 9.7.3   EN 50310

EN 50310: *Application of equipotential bonding and earthing in buildings with information technology equipment* takes the approach that after the safety protective earthing system has been sorted out, reduction of EMC is the next priority. Signal and functional earthing is mentioned but the main detail is reserved to describe how EMC performance will be enhanced by installing a common bonding network (CBN) and MESH-BN (mesh bonding network). The MESH-CBN enlarges the CBN, including the main earthing terminal, by multiple interconnections to the CBN. The mesh system achieves a system reference potential plane (SRPP) which is defined as:

'A conductive solid plane, as an ideal goal in potential equalising, is approached in practice by horizontal or vertical meshes. The mesh width thereof is adapted to the frequency range to be considered. Horizontal and vertical meshes may be interconnected to form a grid structure approximating to a Faraday cage.'

Other statements in this standard include:

- The electrical supply system should be TN-S.
- A reliable signal reference shall be provided by the SRPP.
- A CBN shall be formed by connecting together most metallic items within the building including metallic plumbing, structural steel and protective conductors.
- The CBN may be extended where necessary with a bonding ring conductor which should go alongside the inside perimeter of the equipment room and/or the building. If an optional bonding ring is installed in a room with IT equipment then it shall be directly bonded to the CBN, at least in the four corners of the room.
- A MESH bonding network shall be incorporated within information technology systems. The MESH-BN shall interconnect shelves, cabinets, rack rows, cable racks, ducts, conduits, distribution frames, cable screens and where appropriate a bonding mat.
- The earthing conductor and equipotential bonding conductors shall have insulation according to international and national regulations.

Figure 9.7 shows some of these features.

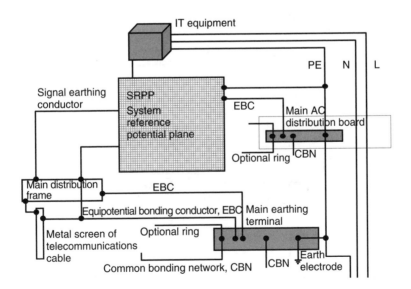

**Fig. 9.7** CBM/MESH-BN network, adapted from EN 50310.

## 9.7.4   ANSI/TIA/EIA-607

In the USA the issue of telecommunications grounding and bonding within commercial buildings is detailed in ANSI/TIA/EIA-607: *Commercial building grounding and bonding requirements for telecommunications.* This standard states its purpose as giving the requirements for:

'A ground reference for telecommunications systems within the telecommunications entrance facility, the telecommunications closet and the equipment room, and

bonding and connecting pathways, cable shields, conductors and hardware at telecommunications closets, equipment rooms and entrance facilities.'

Figure 9.8 shows the elements and disposition of the telecommunications grounding system according to TIA/EA-607.

The main elements of this system are:

- *Telecommunications main grounding busbar (TMGB).* This is the main grounding connection for the whole system and will be located at the telecommunications entrance facility.

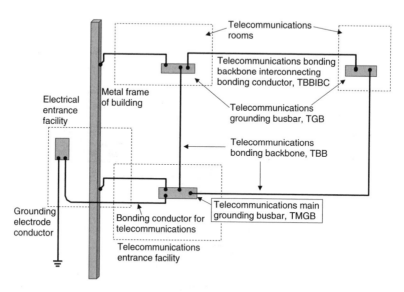

**Fig. 9.8** Grounding system components according to TIA-607.

- *Telecommunications grounding busbar (TGB)*. Every telecommunications space will have its own TGB. Note that busbars are always connected directly to metal frame of the building (where there is one).
- *Telecommunications bonding backbone (TBB)*. All of the grounding busbars are connected together via the TBB.
- *Telecommunications bonding backbone interconnecting bonding conductor (TBBIBC)*. Wherever there may be a long loop of bonding conductors formed by the TBB, equipment and busbars may be connected directly together using a TBBIBC.

Each bonding conductor shall have a conductor size of at least 6 AWG (about 4.1 mm$^2$) although up to 3.0 AWG (about 10 mm$^2$) may be required; it shall be made of copper and be insulated with a distinctive green colour. Every bonding conductor shall be marked with a non-metallic label which shall include the wording '*WARNING If this connector or cable is loose or must be removed please call the building telecommunications manager*'.

The TMGB shall have a size of not less than 6-mm thick, 100-mm wide and be of sufficient length to suit the building requirements. The TGB shall be 6-mm thick, 50-mm wide and of suitable length to meet the building requirements. The busbars must be bonded to the metal frame of the building using 6-AWG conductors. The TMGB shall be 300-mm away from any other cables.

# 9.8  Earthing screened cables

## 9.8.1  What the standards say about earthing screened cables

First, all of the standards agree that that screened cables must be earthed.

- EN 50174 clause 6.3.2 – Screen not bonded to equipment, not recommended.
- En 50310 clause 5.3 – The MESH-BN shall interconnect shelves, cabinets, rack rows, cable racks, ducts, conduits, distribution

frames, *cable screens* and where appropriate a bonding mat to provide the required low impedance of the bonding network.

- BS7671 clause 413-02-2 – To comply with the regulations it is also necessary to apply equipotential bonding to any metallic sheath of a telecommunications cable. (The consent of the owner must be obtained.)

- IEC 60364-4-41 clause 413.1.2.1 – The main equipotential bonding shall be made to any metallic sheath of telecommunication cables. However, the consent of the owners or operators of these cables shall be obtained.

- ANSI/TIA/EIA-568B-1.2 clause 4.6 – The screen of ScTP cables shall be bonded to the TGB in the telecommunication room. Grounding at the work area is usually accomplished through the equipment power connection. Screen connections at the work area shall be accomplished through an ScTP work area patch cord extending from the TO to either the equipment, or to the personal bonding terminal. At the work area end of the horizontal cabling, the voltage measured between the screen and the ground wire of the electrical outlet used to supply power to the workstation shall not exceed 1.0 V rms and shall not exceed 1.0 V DC. The cause of any higher voltage should be removed before using the cable.

## 9.8.2   Effective screen connections

Screened cables must be effectively earthed for the screening elements to be at all effective. Ideally the cable screen would be effectively bonded to earth at both ends for maximum EMC protection, but in many cases this is not possible, with the outlet end terminating inside a plastic box inside a dry plaster wall. It is presumed that the cable screening will find its way to earth via the screened jacks connecting to the screened patch cords which in turn connect through the terminal equipment and into the earth connection of that equipment's mains supply. It is essential therefore to ensure an effective equipotential earthing connection is made to the screened patch panels in the equipment/telecommunications rooms racks. This is the only point where

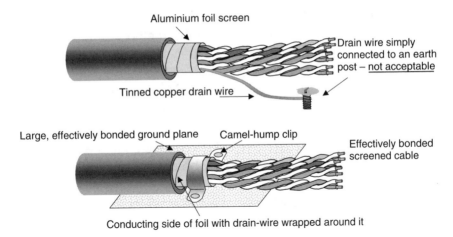

**Fig. 9.9** Effectively bonded screened cable.

an installer can really make any guarantee that at least one high quality bond to earth has been made on the screened cabling.

The screen itself should be 360° bonded to the patch panel earthing mechanism. Simply connecting the drain wire in the cable to an earth terminal is not enough. The whole screening element must be covered by a 360° clamp and clamped down to an earth plate whose surface area is larger than the end cross-sectional area of the cable. Figure 9.9 demonstrates this.

# 9.9   Conclusions

There are many standards around dealing with power distribution and earthing, grounding and bonding. The standards usually approach the subject from totally different positions, being either concerned with the safety aspects of the earth system or the telecommunications/EMC requirements of earthing. This leads to a variety of standards that sometimes appear to offer contradictory advice that is buried under mountains of verbiage where often few solid facts are successfully mined.

However, a few design principles can be determined:

- Design the power distribution system so that it itself minimises electrical noise, e.g. the TN-S method.
- Design the protective earth system so that it does not contribute to noise, e.g. running protective earths close to communications cables and not forming large inductive loops.
- Install FEs where these are required by the customer or the equipment manufacturer.
- Screened cabling must be connected to the main earthing terminal by cables which may be described as FEs or supplementary bonding conductors.
- The protective earth may be used for functional earthing if it is of sufficient quality.
- FEs must not be the only method of protective earthing.
- A common bonding network should be instituted.
- The best EMC performance will be achieved by a mesh structure incorporating conductive mesh in the floors.
- Ensure all screened patch panels have their own functional earth connection that are all individually star-wired back to a telecommunications grounding busbar within every equipment rack via a cream conductor of at least 1.5-mm$^2$ size (about 14 AWG). The TGB must be wired back to the CBN or MESH-CB via a cream cable of at least 4-mm$^2$ (6 AWG) conductor size.

When designing to ISO 11801 2nd edition, it is IEC 60364 that is referred to as the main guidance for earthing systems. In the UK, BS 7671 and BS 6701 must also be obeyed. Within the European Union, where there are no more specific regulations, then EN 50174 and EN 50310 must be followed. In the USA TIA/EIA-607 and TIA/EIA-568-B1.2 must be obeyed.

## 9.10   Full list of relevant standards

- ANSI/NECA/BICSI 568-2001  Standard for Installing Commercial Building Telecommunications Cabling.
- ANSI/TIA/EIA-568-B.1-2  Grounding and Bonding Specifications for Screened Horizontal Cabling.

- ANSI/TIA/EIA-607    Commercial Building Grounding and Bonding Requirements for Telecommunications.
- BS 6701    Code of practice for installation of apparatus intended for connection to certain telecommunications systems.
- BS 7671    Requirements for electrical installations, *also known as the* IEE Wiring Regulations 16th edition *and its corresponding* Guidance Note No. 5 Protection Against Electric Shock.
- EN 41003    Particular safety requirements for equipment to be connected to telecommunication networks.
- EN 50310    Application of equipotential bonding and earthing in buildings with information technology equipment.
- EN 50174-2    Information technology – cabling installation – Part 2: Installation and planning practices inside buildings.
- ETS 300 253    Equipment engineering (EE) – Earthing and bonding of telecommunication equipment in telecommunication centres.
- IEC 60364-1    Electrical installation of buildings – Part 1: Fundamental principles, assessment of general characteristics, definitions.
- IEC 60364-4-41    Electrical installation of buildings – Part 4-41: Protection for safety – Protection against electric shock.
- IEC 60364-5-548    Electrical installation of buildings – Part 5: Selection and erection of electrical equipment – Section 548: Earthing arrangements and equipotential bonding for information technology installations.
- IEC 61024    Protection of structures against lightning.
- IEC 61140    Protection against electric shock.
- ISO 11801 2nd edition    Information technology – cabling for customer premises.

- ISO/IEC TR 14763-2    Information technology – implementation and operation of customer premises – Part 2: Planning and installation.
- ITU-T K.27:1996    Bonding configurations and earthing inside a telecommunication building.
- ITU-T K.31:1993    Bonding configurations and earthing of telecommunication installations inside a subscriber's building.
- ITU-T Handbook    Earthing of telecommunication installations (Geneva 1976).

# 10

# Administration schemes

## 10.1 Introduction

For a structured cabling system to maintain its value as an asset it is necessary to maintain accurate records of what the cabling consists of, what kind of connections there are, what kind of numbering scheme is employed and to have an organised method of administering changes. All this information collected together is known as the administration scheme.

The method of recording this information may be varied. There may be nothing at all; the information may be handwritten on a scrap of paper; it may be neatly recorded in a workbook; it may be listed in a Microsoft Excel spreadsheet or some other common database package like Microsoft Access. There are specially written visually oriented databases designed specifically to record structured cabling systems and at the top of the evolutionary chain are intelligent patch panel systems linked to real-time dedicated databases. Whichever method is used there is a common set of philosophies described in various standards that attempt to lead the user into a logical and unified recording structure. These standards are:

- ISO/IEC 14763-1: *Information technology – Implementation and operation of customer premises cabling – Part 1: Administration.*
- EN 50174-1: *Information technology – cabling installation – Part 1: Specification and quality assurance.*

- ANSI/TIA/EIA-606: *Administration Standard for the Telecommunications Infrastructure of Commercial Buildings.*

## 10.2   Definitions (adapted from IEC 14763)

- *Identifier.* A unique item of information that enables a specific component of the information technology infrastructure to be differentiated in the administration records.
- *Label.* A label is used to mark clearly a specific component of the information technology infrastructure with its identifier and (optionally) other information.
- *Pathway.* Cable route (e.g. conduit, ductwork, tray or tube) used to accommodate cables between termination points defined by a physical structure.
- *Record.* A collection of information about or related to a specific element of the information technology infrastructure.
- *Space.* An area (e.g. closet, cabinet, manhole or equipment room) used to house cable terminations or equipment.
- *Work order.* A collection of information which documents the changes requested and the operations to be carried out on the information technology infrastructure.

## 10.3   Administration philosophy

The items for which we must keep records and identifiers are:

- pathways.
- spaces.
- cables.
- terminations.
- grounding.

Although IEC 14763 does not mention the terms, most commercial database packages that target structured cabling focus on libraries and inventories.

The *library* defines all the possible items (and possibly locations) that are available to the user. This could tie items down to part numbers, manufacturers and so on. It can incorporate all the hardware of the cabling system plus the first levels of the active information technology equipment such as the LAN switches and PABXs.

The *inventory* lists exactly what the user does have and where it is and so a *locations* database must also be included. The connectivity database can then list what the connections are and where to find them. A unique numbering code is then assigned to all the items that need to be recorded. The list above shows that the pathways and grounding points are also considered part of the administration scheme.

The outputs from the scheme are *reports* and *work orders*. The reports can be configured to give the user whatever information he or she thinks is necessary from the database. A work order is a method of raising a change request and subsequently allows that change to be documented within the recording system or database. Figure 10.1 shows all the elements of a cabling administration scheme.

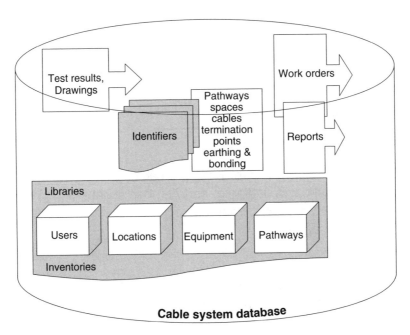

**Fig. 10.1** Cabling administration scheme.

## 10.4   Records

IEC 14763 suggest the following items should be recorded, although apart from the minimum records requirement this consists of more information than most companies would need or have time to enter into a database.

### 10.4.1   Minimum records

The following minimum records regarding cabling infrastructure shall be provided:

- for cables: locations of end points, type, number, pairs;
- for outlets: identifier, type, location;
- for distributors: identifier, designation, type, location, connections;
- the floor plan, including the locations of the outlets, distributors and pathways.

### 10.4.2   Optional records

When changes are made to the cabling infrastructure, including pathways and spaces, additional records may be necessary.

### Cable records

Cable records describe:

- type of optical fibre or copper cable
- typical cable data (e.g. part number, sheath colour)
- sheath and core identification
- manufacturer
- number of unterminated conductors and those with failures
- length
- data such as attenuation and crosstalk
- identification of pin connections at both ends and of splices
- performance classification (if applicable)
- location of earthing

- treatment of screens
- transmission system under operation
- date code
- part number
- identifier
- linkage to identifiers for distributors, outlets, pathways and spaces.

## Telecommunications outlet records

TO records describe:

- performance classification (if applicable)
- single mode or multimode fibre (50/125 or 62.5/125)
- screened or unscreened design
- manufacturer
- number and arrangement of terminated pins if not all pins are terminated
- part number
- identification of ports and cables connected
- linkage to identifiers for distributors, outlets, pathways and spaces.

## Distributor records

Distributor records describe:

- number of available and used cables, fibres or pairs
- manufacturer
- number of conductors
- linkage to identifiers for cables, pathways and spaces
- part number
- front view of the patch cabinet.

## Pathway records

Pathway records describe:

- type
- metal or non-metal design

- dimensions, mechanical data
- branching points
- manufacturer
- identification
- length
- location
- records of cables installed in that pathway
- location of earthing.

## Space records

Space records describe:

- locations
- dimensions
- identification
- equipment located in the spaces
- space available
- type.

## Active components records

Active components records describe:

- type of device
- model number
- availability of cables (number of ports)
- identifier (MAC or IP address)
- adaptation of ports
- identification of ports
- location of device
- manufacturer
- name of user, department, telephone extension
- location of telecommunications outlets
- serial number, date of installation.

## *Other information*

Details of protocols may also be recorded along with drawings, work orders and results of link and channel measurements.

## 10.5   Database formats

IEC 14763 recommends a five-field database format, where:

- Field 1 is the general location.
- Field 2 is the specific location.
- Field 3 is the component identifier.
- Field 4 is the port number.
- Field 5 is physical data.

The standard recommends a minimum of five digits for the first field, seven for the second and four for the third. There is no comment on the size of fields 4 and 5.

Database construction for cabling administration is mostly a matter of common sense. People will want to know where an outlet is, where the other end is and what kind of service is available from that communications link. Figure 10.2 gives an example of a paper-based cable record.

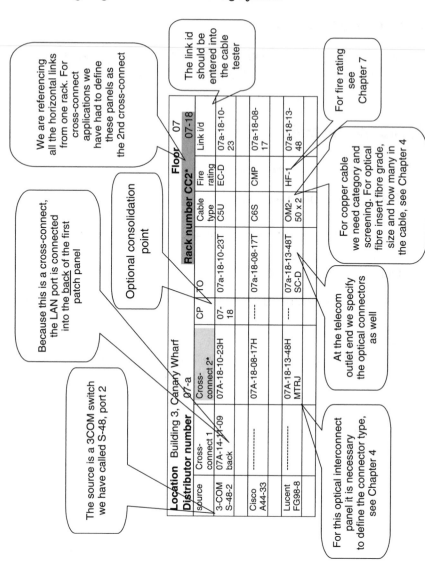

**Fig. 10.2** Example of a cable record sheet.

## 10.6    Colour coding

TIA/EIA-606 is the only administration standard which remarks upon colour coding, apart from the multimode/singlemode differentiation commented upon in ISO 11801 2nd edition. Table 10.1 gives the TIA/EIA-606 colour coding scheme. Some people have interpreted this colour scheme to mean the actual sheath colour of the cables themselves, for example, blue is given as the identifier for the station telecommunications media, that is, the horizontal cabling. This is very

Table 10.1  TIA/EIA-606 colour coding scheme

| Termination type | Colour | Pantone colour code | Comments |
| --- | --- | --- | --- |
| Demarcation point | Orange | 150C | Marks the point of demarcation between owners or responsible parties, usually in the entrance facility |
| Network Connections | Green | 353C | Used to identify the termination of network connections on the customer side of the demarcation point |
| Common equipment | Purple | 264C | Major switching and datacommunications equipment, PBX, LANs, etc. |
| First level backbone | White | | Main cross-connect to intermediate cross-connect terminations |
| Second level backbone | Grey | 422C | Intermediate cross-connect to telecommunications closet connection |
| Station | Blue | 291C | Horizontal cable terminations. Only required at the telecommunications and equipment room end, not at the telecommunications outlet |
| Interbuilding Backbone | Brown | 465C | Campus cable terminations |
| Miscellaneous | Yellow | 101C | Auxiliary circuits such as alarms, maintenance, security, etc. |
| Key telephone systems | Red | 184C | |

Termination labels identifying the two ends of the same cable shall be the same colour. Cross-connections are generally made between termination fields of two different colours. Labels may be marked with the category of cable.

much a minority interpretation and the vast majority of all the horizontal cabling shipped in the world today remains grey coloured, albeit every shade of grey possible.

The colour coding is to be used for labelling at the cable ends and includes cable labels, coloured patch cords and coloured inserts in patch panels and outlet plates.

## 10.7   Intelligent patch panel systems

The year 2002 saw the introduction of intelligent patch panel systems from no less than six manufacturers, with no doubt more to follow. It is therefore worth considering what intelligent patching exactly means, what it can do and where it might be used economically.

There are already graphically based computerised database packages available that are designed to record the cabling records. They are acknowledged in the standards as being very desirable for the larger projects.

### EN 50174-1 clause 8.1

For simple infrastructures a well-designed paper-based administration system is adequate. However, it is recommended that the principles of administration outlined in this clause be implemented using a computer-based system.

### IEC 14763-1 clause 4.2

It is recommended that the principles of administration outlined in this clause be implemented using a computer-based administration system. For smaller, less complex systems, a well-designed paper-based administration system may be adequate. The complexity of the administration system may be related to the size of the telecommunications infrastructure. For a small system, a customised commercial database programme may be adequate. For a large organisation, the cabling administration system may require a sophisticated database, an efficient data retrieval program and additional features. For example, the computer administration package may input drawings directly from CAD programs or may output reports to external packages or e-mail work orders and automatically update records on completion of work and may also serve as a cabling design tool.

The essential difference with an intelligent patching system is that the patch panels can detect what they are connected to and can maintain and update the database in real time, automatically. Every manual off-line cabling database is destined to lose value with time as people using it fail to input the changes made manually.

The systems currently available all use the cross-connect method. One patch panel is dedicated to the equipment and one to the cabling. The patch panels are constructed so that they can tell which port is connected to the corresponding port on the panel. One method of doing this uses a patch cord with an extra wire in it. There is thus a hardwired extra link between panels to identify the port connectivity.

It is all very well having a patch panel that knows what it is connected to but for the system supervisor to be able to access that information an extra item is required. This is known as a scanner. It is a box of electronics that sits in each equipment rack and is connected to each of the intelligent patch panels. The scanner polls each panel in turn to find out what it is connected to. The scanner is linked in to the corporate LAN to relay that information back to the supervisory position. Somewhere on the company server sits the overall database management package that oversees the operation and presents the information in a graphical manner to the user. Figure 10.3 shows an example of such a system, based on the Brand-Rex SMARTPpatch™ system.

(a)

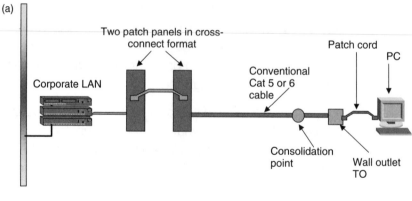

(b)

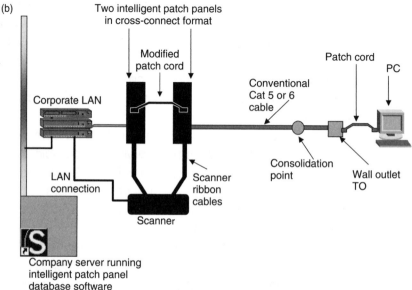

**Fig. 10.3** (a) Conventional horizontal cable system. (b) Horizontal cable system with intelligent patching components.

In such a managed cabling system the horizontal cable, CP, TO and patch cords at each end of the channel are conventional Cat 5, Cat 6 or optical fibre elements. The unique items are the two intelligent patch panels forming the cross-connect and the special patch cord between them containing the extra wire. The scanner is connected into each panel via a ribbon cable. The scanners are connected back to a master scanner which in turn is connected into the corporate LAN, either locally or via telecommunications or internet connections. In the company server sits an SQL based database package with a graphic front-end called SP4E, or Smartpatch for the Enterprise.

Any change that is made to any of the connectivity is reported back to the supervisory position, wherever it may be, as soon as the polling routine has scanned that panel and reported back the change.

Work orders may be generated by the supervisor pulling up a picture of the link and making the necessary changes on screen. A technician can access this work order from any other screen and go to the equipment rack in question. Here flashing LEDs over the appropriate ports will flash to guide him or her into making the right changes. As soon as a change is made, correct or incorrect, a message will appear at the supervisory screen informing the manager of the change.

The cabling database is automatically upgraded whenever a change is made, so it is always correct. The SMARTPatch system will also interrogate the terminal devices to see if they have a MAC or IP address, and add these devices to the database as well.

Needless to say, intelligent patching is not cheap, increasing the hardware pricing typically by over 100%, so it is worth considering where such applications would be cost effective.

There is a hidden cost of ownership when using structured cabling. When a PC is reported to be faulty somebody from the IT department will have to go to that office, see where the connection is made and then follow that link within the telecommunications room to see what is plugged into what. With a chaotic patching field with no records this can take hours. If it is a remote site it can take days and include a trip from head office to the remote site. With a large installation and/or a very dispersed one, the hidden cost of ownership can be very large. Intelligent patching can:

- Reduce the costs of network downtime.
- Reduce the costs of administering moves adds and changes.
- Reduce the skill level requirements of many of the IT technicians required to maintain a large system.

Cost models from manufacturers suggest that projects with more than about 2000 outlets and/or more than six sites to manage can benefit financially from intelligent patching within a three-year pay-back timescale.

# 11

# Testing

## 11.1　Introduction

Cable systems have to be tested after installation to prove that they meet the standard they were procured against. If they are not tested, or not tested correctly, then it is a gamble whether they will work or not and whether they do will be at the user's own risk. Testing telephone cable networks used to involve little more than a continuity check for each conductor. For undemanding applications not much more was needed, but with the advent of Category 6, optical fibre and multi-gigabit transmission systems most people would agree that thorough testing of the cable plant is essential.

Testing standards came along after the first wave of cabling system standards and for many years the only standard available was TSB 67, the Telecommunications Systems Bulletin number 67 of TIA/EIA-568. Now we have many standards relating to testing and some challenging testing environments for Category 6 cabling, single mode optical fibres and a new generation of optical connectors.

Contractually, testing is a good idea so that installer and user can have an unambiguous set of parameters to determine when the cable system is working satisfactorily, so that it can be handed over and monies paid.

## 11.2   Testing standards

### 11.2.1   ISO 11801 2nd edition

ISO 11801 2nd edition refers to the following standards:

- IEC 61935-1        Generic cabling systems – Specification for the testing of balanced communication cabling in accordance with ISO/IEC 11801 – Part 1: Installed cabling.
- IEC 61280-1-1      Fibre optic communication subsystem basic test procedures – Part 1.1: Test procedures for general communications subsystems – Transmitter output optical power measurement for single mode optical fibre cable.
- IEC 61280-4-1      Fibre optic communication subsystem basic test procedures – Part 4.1: Test procedures for fibre optic cable plant – Multimode fibre optic plant attenuation measurement.
- IEC 61280-4-2      Fibre optic communication subsystem basic test procedures – Part 4.2: Test procedures for fibre optic cable plant – Single mode fibre optic plant attenuation measurement.
- IEC 61280-4-3      Fibre optic communication subsystem basic test procedures – Part 4.3: Test procedures for fibre optic cable plant – Single mode fibre optic plant optical return loss measurement.
- ISO/IEC 14763-3    TRT3 Information technology – Implementation and operation of customer premises cabling – Part 3: Testing of optical fibre cabling.

### 11.2.2   EN 50173

EN 50173 refers to the following as the principal test standard:

- EN 50346          Information technology – Cabling installation – Testing of installed cabling.

EN 50346 in turn makes reference to the following:

- EN 61280-4-2    Fibre optic communication subsystem basic test procedures – Part 4–2: Fibre optic cable plant – Single mode fibre optic cable plant attenuation (IEC 61280-4-2:1999).
- EN 61300-3-4    Fibre optic interconnecting devices and passive components – Basic test and measurement procedures – Part 3–4: Examinations and measurements – Attenuation (IEC 61300-3-4:1998).
- EN 61300-3-6    Fibre optic interconnecting devices and passive components – Basic test and measurement procedures – Part 3–6: Examinations and measurements – Return loss (IEC 61300-3-6:1997).
- EN 61300-3-34   Fibre optic interconnecting devices and passive components – Basic test and measurement procedures – Part 3–34: Examinations and measurements – Attenuation of random mated connectors (IEC 61300-3-34:1997).
- EN 61935-1      Generic cabling systems – Specification for the testing of balanced communications cabling in accordance with EN 50173 – Part 1: Installed cabling (IEC 61935-1:2000).

## 11.2.3   ANSI/TIA/EIA-568-B

TIA/EIA-568-B is a bit more self-contained with detail of test requirements contained within the standard. There are some references to optical system testing, however:

- ANSI/TIA/EIA-526-7       Optical Power Loss Measurements of Installed Single mode Fiber Cable Plant.
- ANSI/TIA/EIA-526-14-A    Optical Power Loss Measurements of Installed Multimode Fiber Cable Plant.

# 11.3  Copper cable testing

## 11.3.1  Required tests

IEC 61935-1 requires the following tests:

- Wire map, including cable screen if present.
- Attenuation (called insertion loss in ISO 11801 2nd edition).
- NEXT, pair to pair, local and remote.
- Power sum NEXT, local and remote.
- ELFEXT, pair to pair.
- Power sum ELFEXT, pair to pair.
- Return loss, local and remote.
- Propagation delay.
- Delay skew.
- DC loop resistance.
- Length (not listed as a required test but described in the text as part of the 'inspection of workmanship' requirements).

## 11.3.2  What the tests actually mean

### Wire map

This is a continuity test that ensures that there are no open or short circuits. Every conductor goes to the right connector pin at the far end and there are no reversed pairs, transposed pairs or split pairs. Figure 11.1 demonstrates these conditions. If there is a cable screen present its continuity is also checked.

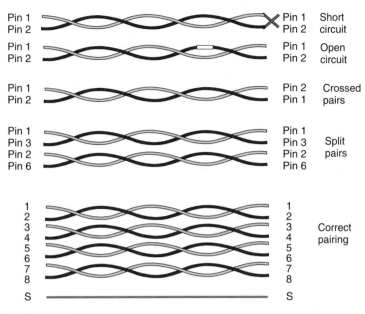

**Fig. 11.1** Wire map.

## *Attenuation*

This is now called 'insertion loss' in ISO 11801 2nd edition to take into account other losses in the cable apart from just attenuation, for example return loss. It means how much energy in the signal has been absorbed as it travels down the cable. It is frequency and length dependent. The unit of measurement is the decibel, dB.

## *NEXT*

NEXT means near end crosstalk. This is the amount of signal energy leaking from one pair into any adjacent pair. With four pairs there are six possible combinations. It is called near-end because it is measured right next to the transmitter where the crosstalk is going to be strongest. NEXT is the largest contributor to noise within a structured cabling system. NEXT is frequency dependent and the unit of measurement is the decibel. The test regime must test this parameter at both ends of the link.

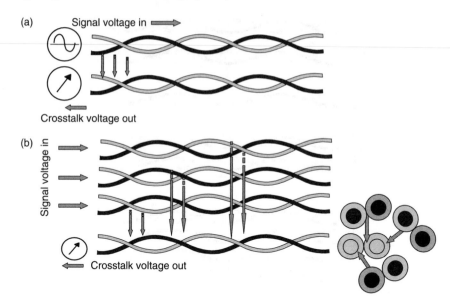

(a) Signal voltage in ⟹

Crosstalk voltage out

(b) Signal voltage in

Crosstalk voltage out

**Fig. 11.2** (a) Near end crosstalk – pair-to-pair. (b) Near end crosstalk – Powersum.

When the insertion loss of the cabling of Class D, E or F is below 4.0 dB then the NEXT requirement is not applicable.

## *Power sum near end crosstalk, PS-NEXT*

NEXT, on its own, is a measurement of the pickup from one pair to another. In a real life cable, however, there are at least four pairs and each one might be carrying a signal, for example in 1000BASE-T applications. The real crosstalk suffered by any one pair is therefore a combination of all the pair-to-pair NEXT values contributed by the three other pairs surrounding it. A more demanding test is therefore PS-NEXT. PS-NEXT is frequency dependent and the unit of measurement is the decibel. The test regime must test this parameter at both ends of the link. Figure 11.2 shows NEXT and PS-NEXT.

When the insertion loss of the cabling of Class D, E or F is below 4.0 dB, the PS-NEXT requirement is not applicable.

## *ELFEXT*

ELFEXT means equal level far end crosstalk. First we have to define far end crosstalk (FEXT) which is the amount of signal leaking from one pair to the adjacent pair when measured at the other end of the cable away from the transmitter. It used to be assumed that NEXT and FEXT performance would be the same, but practical measurements have shown this not to be the case. FEXT has become important because of protocols such as 1000BASE-T utilising simultaneous bidirectional transmission down each pair, and the fact that FEXT is much harder to cancel out electronically than NEXT.

Sophisticated transmitter circuitry can accommodate a certain amount of NEXT. If each transmitter sends a known signal in turn onto each of the pairs to which it is connected, and, at the same time the receiver circuits connected to the three adjacent pairs 'listen' for any signal appearing at the same time on those adjacent pairs, then it can reasonably be assumed that they are detecting the crosstalk between those pairs. If, for future transmissions, a suitable 'inverse' signal is applied to the other receiving pairs, a large part of the NEXT can be electronically cancelled out.

FEXT is much harder to accommodate. First, the effect is appearing about 500 ns later at the far end of the cable and is mixed with other crosstalk and external noise components. Second, what effect will any compensation for FEXT have on the NEXT performance?

Thus FEXT, along with external noise, is extremely difficult to cancel out by digital signal processing, see Fig. 11.3.

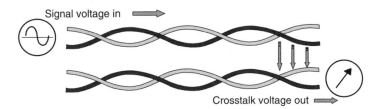

Signal voltage in ⟹

Crosstalk voltage out ⟹

**Fig. 11.3** Far end crosstalk – pair-to-pair.

The effect of FEXT can be masked by the presence of the attenuation of the cable. The latest structured cabling standards thus require that ELFEXT be measured and reported. ELFEXT is the difference between the FEXT and the attenuation of that link. By putting the attenuation of the cable back into the equation it gives a length-independent normalised value which is of more use than FEXT on its own.

## *PS-ELFEXT*

As with Power Sum NEXT, which is a more representative function of the overall NEXT of a cable, ELFEXT is complemented by power sum ELFEXT. Power sum ELFEXT (PS-ELFEXT) is the sum of the combined effect of all the FEXT of all the other pairs in the cable, less the attenuation of the victim pair. With NEXT, in a four-pair cable, there will be six possible combinations; power sum measurements will have four combinations. ELFEXT and PS-ELFEXT need only to be measured from one end although the cable tester used may test from both ends as a matter of routine. Once again the unit of measurement is the decibel.

## *Return loss*

Return loss is the amount of energy reflected back from a circuit owing to an impedance mismatch between the source, the cable, the load, or all three. If the input impedance of the source matches the characteristic impedance of the cable, which in turn is terminated by a load of the same impedance, then all the signal energy, minus that lost because of attenuation, will be transferred to the load. Return loss is another unwanted loss of signal energy, but more importantly, out-of-phase signals reflected back into the transmitter could create havoc with modern multilevel, multiphase coding schemes.

Load matching would be ideal but is not practical in the real world. For example, the 100-$\Omega$ cable normally specified in LANs and the cabling standards is allowed in reality a characteristic impedance of $100 \pm 15\,\Omega$ and will still be within specification.

Instead of measuring the characteristic or input impedance of the circuit, which is difficult, it is considered more practical and useful to measure the result of impedance mismatches, that is, return loss, see Fig. 11.4.

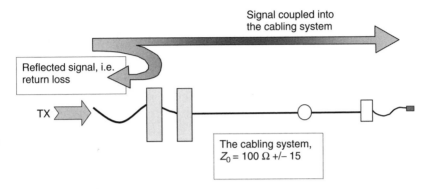

**Fig. 11.4** Return loss.

Excessive return loss may be seen from short links caused by the addition of the return loss from the far end before this is, in turn, attenuated by the cable. This reflection can give misleading results and therefore ISO 11801 2nd edition allows for all return loss readings at frequencies where the insertion loss is below 3 dB and can be ignored. This would equate to approximately 15 m of Category 5 cable. The exact length will depend upon the frequency of operation considered and the grade of cable used. The higher the grade or category of cable, the lower the attenuation. Similarly the higher the frequency, the higher is the attenuation. This '3-dB' rule should already be written into the software of the various commercially available cable testers on the market.

## Propagation delay and delay skew

The length of time taken for the signal to transit the entire cable system is known as the propagation delay. The latest standards call for a delay of no more than around 570 ns to transit the entire length of a 100-m cabling system. Some LANs such as Ethernet are sensitive to time delay in the transmission medium.

A four-pair cable consists of four communications channels, each having a slightly different delay time. The difference in delay is caused by the differing lay length on each pair (thus giving a different physical length of each pair) and possibly by the use of different insulation materials for each pair which gives a different nominal velocity of propagation (NVP) for each pair. LAN protocols that share the data across all four pairs such as 100BASE-T4 and 1000BASE-T are very sensitive to differential delay.

The terms used may be skew, asymmetric skew or differential delay. The standards call for no more than a 45–50-ns delay between the 'fastest' and 'slowest' pair in a cable system. Excessive delay skew will also upset RGB (red, green, blue) video systems where the red signal component travels on one pair, green on another and blue on a third. Delay skew will cause colour fringing at the receiving end which can only be cured with expensive delay lines.

## DC loop resistance

The direct current (DC) resistance is a major contributor to the overall attenuation and must be minimised. 'Loop' resistance means the resistance of the combined pair as if one end had been shorted together and so 1 km of a twisted pair actually represents one wire of 2 km length to the test instrument. The hand-held cable tester will be comparing the measured loop resistance against its expectations for a 90-m link or a 100-m channel.

### 11.3.3   Other copper cable test parameters and issues

## Attenuation to crosstalk ratio

One parameter that is often discussed, presented in manufacturers' data sheets and even specified in the standards is the attenuation to crosstalk ratio (ACR); yet surprisingly this is not a required test. Installers will find, however, that nearly all hand-held testers will always present this information anyway as the default setting.

The ACR is similar in concept, although not identical, to signal-to-noise ratio (SNR). It is a good measure of the overall quality of a cabling link. The higher the ACR the better as it implies that the desired signal is not being so severely attenuated that the effect of crosstalk noise will become too significant and drown out the desired signal. In communications channels it is generally considered that a positive value of ACR is required for successful, error-free transmission. ACR is easily calculated from the formula:

ACR(dB) = NEXT(dB) − attenuation(dB)

The graph in Fig. 11.5 shows how attenuation increases as the frequency increases. The NEXT separation also declines as the frequency increases. The gap between the two functions, the ACR, thus gets less and less as the frequency increases. At some point the two lines will cross over and this will be the point of zero ACR. The zone beyond this line is known as the negative ACR region.

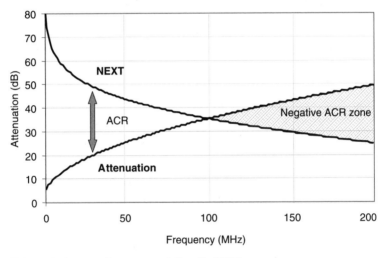

**Fig. 11.5** Attentuation to crosstalk ratio (ACR).

Power sum ACR is the power sum NEXT value of the victim pair minus the attenuation of that pair. A fundamental definition of what Classes A to F and Categories 3 to 7 mean is at what frequency does the PS-ACR go to zero. For Class D, for example, it must be at least 100 MHz. This essentially gives the useable frequency range of the cabling system and is analogous, but not identical, to the system bandwidth.

## Alien crosstalk

Alien crosstalk is defined in many LAN standards but not specified in ISO 11801 2nd edition. Whereas NEXT is crosstalk between pairs in the same cable, alien crosstalk is interference from one cable to an adjacent cable. No on-site test is specified for it and it must be coped with by the design of the cable and the cabling plant.

A cable can only be surrounded by six other cables of the same size, so the size of the cable bundle is not of great significance in determining alien crosstalk. The phenomenon is dependent upon the balance of each cable and the length of parallel cable runs. The cable must be of sufficient quality that it can be installed in a parallel cable run up to 90 m with no significant alien crosstalk pickup problem.

## Nominal velocity of propagation

The speed that the signal travels down the copper cable compared to the speed of light in a vacuum is called the nominal velocity of propagation (NVP). For most data cables it is in the range 0.69–0.75. If the speed of light is 300 000 km/s and if the NVP of a cable is 0.7, the speed of the signal in the cable is $0.7 \times 300 000$ or 210 000 km/s. The value of the NVP is an important factor to know to load into test instruments so that the correct length of cable can be determined.

The only way that the cable tester can measure the length of the cable is to send a pulse of electricity down the cable and wait for it to be reflected back from the far end. If the instrument times this round journey and it knows the speed of the signal (from the NVP), it can calculate the cable length. This is known as time domain reflectometry. The installer must therefore load the correct NVP, as supplied by the cable manufacturer, into the test instrument before the test starts.

## Remote power feeding over the cabling

If it is the intention to use the LAN to distribute the telephone system (i.e. voice over IP) or to supply wireless LAN base stations, then the telephones and base stations must get their electrical power via the structured cabling. Although there is no on-site test to prove the cables' capacity in this respect the cable design must conform to the requirements of the standards.

ISO 11801 2nd edition states an electrical power distribution requirement for the copper cables to be used in the horizontal, be they Category 5, 6 or 7:

- Voltage capacity: 72 V.
- Current carrying capacity: 0.175 A DC per conductor.
- Power capacity: 10 W per pair.

This is exactly the same as the cable specification associated with ISO 11801 2nd edition, namely IEC 61156 *Symmetrical pair/quad cables for digital communications.*

A new IEE standard covering remotely powered Ethernet is IEEE 802.3af, encompassing 10BASE-T, 100BASE-T and 1000BASE-T. This standard defines six classes of remote power feeding with power levels up to 15 W at the input and 12.95 W at the output end. This is slightly higher than the ISO 11801 2nd edition individual pair requirement but poses no problems for good quality cabling.

## *Characteristic impedance*

Characteristic impedance is very important because mismatched cable elements will reflect signal energy back to the source and be seen as return loss. Structured cabling components need to be manufactured with a characteristic impedance of 100 $\Omega$ plus or minus 15 $\Omega$. Characteristic impedance is very difficult to measure with a simple hand-held test instrument and the result is only going to give the average input impedance 'seen' from one end of the cable. For this reason it is important to measure the practical effect of characteristic impedance mismatch, that is return loss. So although the tester will measure and report characteristic impedance it should only be viewed as 'for information only' and not be seen as part of the acceptance criteria.

## The effects of temperature on copper cabling

It must be understood that copper cable performance declines as the temperature goes up. This fact of life is acknowledged in the standards but ignored by the hand-held test instruments which have to assume that the whole world is constantly at 20°C because that is what the standard channel and link parameters are normalised to.

ISO 11801 2nd edition de-rates horizontal cables lengths as the temperature rises because of the increase in attenuation that the cable will suffer. Above 20°C, the transmission distance should be reduced by 0.2% per °C for screened cables and 0.4% per °C for unscreened cables.

## Category 6 interfaces

When testing a Class D/Category 5 link with one of the commercially available hand-held testers available today, it is necessary to put on the input of the device a generic Category 5 interface. With Class E/Category 6 the market is moving towards generic interfaces but at the time of writing this has not yet happened. When testing Category 6 cabling installations it is essential to check with the cable manufacturer which interface they approve or recommend for each type of tester. Failure to do this will allow the manufacturer to abrogate any responsibility for the cabling plant performance.

## The test report and saving the test results

The test instrument must be capable of recording the following information as detailed in Table 11.1.

After each test with the hand-held tester the operator has the option of saving the test results. This may be essential as part of the contractual hand-over documentation or as a required action to receive a manufacturer's warranty.

The test results must be saved in a file format that can be understood at a later date. The tester will usually allow the files to be saved in a number of computer formats, for example .txt, .xls and so on. But they can only be manipulated if they are saved in the same format

Table 11.1 Summary of reporting requirements for field test equipment

| Function | Measurement (Single ended test unless otherwise stated) | |
| --- | --- | --- |
| **Wire map** | All connectivity including screens if present | Pass/Fail |
| **Attenuation** (insertion loss) | Worst case attenuation (of the four pairs) Test limit at worst case attenuation Frequency at worst case attenuation Wire pair with worst case attenuation | Pass/Fail |
| **NEXT pair-to-pair** This must be measured at both local and remote ends and the worst cases reported | Worst case pair-to-pair NEXT loss (1 of 6 possible) Test limit at worst case NEXT Frequency at worst case NEXT Pair combination with the worst case NEXT Worst case pair-to-pair NEXT margin (1 of 6 possible) Test limit at worst case NEXT margin Frequency at worst case NEXT margin Pair combination at worst case NEXT margin | Pass/Fail |
| **PS-NEXT** This must be measured at both local and remote ends and the worst cases reported | Worst case pair-to-pair PS NEXT loss (1 of 4 possible) Test limit at worst case PS NEXT Frequency at worst case PS NEXT Pair combination with the worst case PS NEXT Worst case pair-to-pair PS NEXT margin (1 of 4 possible) Test limit at worst case PS NEXT margin Frequency at worst case PS NEXT margin Pair combination at worst case PS NEXT margin | Pass/Fail |
| **ELFEXT pair-to-pair** | Worst case pair-to-pair ELFEXT loss Test limit at worst case ELFEXT Frequency at worst case ELFEXT Pair combination with the worst case ELFEXT Worst case pair-to-pair ELFEXT margin Test limit at worst case ELFEXT margin Frequency at worst case ELFEXT margin Pair combination at worst case ELFEXT margin | Pass/Fail |
| **Power sum ELFEXT** | Worst case PS ELFEXT loss Test limit at worst case PS ELFEXT | |

Table 11.1 *Continued*

| Function | Measurement (Single ended test unless otherwise stated) | |
|---|---|---|
| | Frequency at worst case PS ELFEXT | |
| | Pair combination with the worst case PS ELFEXT | |
| | Worst case PS ELFEXT margin | |
| | Test limit at worst case PS ELFEXT margin | |
| | Frequency at worst case PS ELFEXT margin | |
| | Pair combination at worst case PS ELFEXT margin | Pass/Fail |
| **Return loss** This must be measured at both local and remote ends and the worst cases reported | Worst case return loss margin (1 of 4 possible) Test limit at worst case return loss margin Return loss at worst case RL margin Frequency at worst case RL margin | Pass/Fail |
| **Propagation delay** | Worst case propagation delay (1 of 4 possible) Test limit at worst case propagation delay | Pass/Fail |
| **Delay skew** | Worst case delay skew Test limit of delay skew | Pass/Fail |
| **DC loop resistance** | Worst case DC loop resistance (1of 4 possible) Test limit of DC loop resistance | Pass/Fail |

as that manufacturer's own cable software management programme. For example, if using a Fluke tester then the file must be stored in .fcm format as to be compatible with Fluke Cable Manager™.

## 11.3.4   Test accuracy and marginal test results

It has to be remembered that hand-held testers are not laboratory instruments but workman's tools that spend most of their lives bouncing about in the back of a white van. There is a great temptation to accept digital readouts from machines at face value but such values come with a tolerance and margin of error. This is recognised

Table 11.2 Estimated measurement accuracy at the channel pass/fail limit for level IIE test instruments from IEC 61935-1

| Test parameter | Baseline accuracy | Accuracy with test adapter |
|---|---|---|
| Attenuation | 1.3 dB | 1.9 dB |
| Pair-to-pair NEXT | 1.8 dB | 3.6 dB |
| Power sum NEXT | 1.8 dB | 3.9 dB |
| Pair-to-pair ELFEXT | 2.4 dB | 4.4 dB |
| Power sum ELFEXT | 2.5 dB | 4.7 dB |
| Return loss | 1.7 dB | 2.7 dB |
| Propagation delay | 25 ns | 25 ns |
| Delay skew | 10 ns | 10 ns |
| Length | 5 m | 5 m |
| DC resistance | 1.4 Ω | 1.4 Ω |

in standards such as IEC 61935 and TIA/EIA-568-B. Table 11.2 gives the estimated measurement accuracy at the channel pass/fail limit for level IIE test instruments at 100 MHz.

Level IIE is a third generation test requirement to coincide with Cat5e, obviously superseding level I and II, but to be surpassed by level III for Class E. Test equipment manufacturers will claim much better performance for their own products but how a machine performs in the factory, after calibration at 20°C, may not be the same after it has been in a tool box in the back of the white van for two years.

It is a good idea for installer and user to agree, in advance, as part of the contractual terms how failures, and in particular marginal failures, will be dealt with. EN 50346 makes the following statement:

'Marginal results may be treated in a number of ways including:
(a)   verification of the normalisation of the test system
(b)   acceptance of all marginal results
(c)   rejection of all marginal results
(d)   acceptance of marginal pass results and rejection of marginal fail results
(e)   repetition of the measurement using a test system with an improved measurement accuracy.

*It is recommended that the approach to be adopted be agreed before the testing is undertaken.'*

It is also a good idea to understand the manufacturer's approach to marginal failures when a manufacturer's warranty is expected for the cabling system.

## 11.3.5   EN 50346

EN 50346 is the test Standard to accompany EN 50173. It is similar to IEC 61935 except that it includes optical fibre tests and does not lay down test requirements, only possible tests that should be defined in the contract stage.

'This standard does not define which tests should be applied or the quantity or percentage of installed cabling to be tested. The test parameters to be measured and the sampling levels to be applied for a particular installation should be defined in the installation specification and quality plans for that installation prepared in accordance with EN 50174-1.'

The test regime fully described in this Standard covers the following:

- Wire map (short to screen, short circuit, screen continuity).
- Length.
- Propagation delay.
- Delay skew.
- Attenuation (insertion loss).
- Attenuation (insertion loss) deviation.
- Near end crosstalk loss (NEXT, pair-to-pair).
- Power sum NEXT.
- Equal level far end crosstalk loss (ELFEXT, pair-to-pair).
- Power sum ELFEXT.
- ACR (pair-to-pair).
- Power sum ACR.
- Return loss.
- Unbalance attenuation, near end (LCL).
- Coupling attenuation.

- DC loop resistance.
- Resistance unbalance.

## 11.3.6  Copper cable testing: summary

To test the link, use one of the recognised hand-held cable testers. Follow the manufacturers' instructions and ensure that the test leads used are the correct ones (they should be the same leads used for each test and kept exclusively for testing purposes). Request from the manufacturer the latest software release for the machine that you are using.

The particular model chosen must be specifically designed to test installed cable to Category 5 or 6 standards. Older versions of cable testers that can only measure Category 5 parameters to the original 1995 specification are known as Level II. To measure the new enhanced or Cat5e parameters Level IIE testers are required. Category 6/Class E measurements require a Level III tester.

- Agree with the user what percentage of the installation is to be tested.
- Agree with the user how the test results are to be presented.
- Agree with the user how marginal fails will be handled.
- All four pairs must be terminated at both ends of the cable and all four pairs tested.
- The cable must be tested in both directions. It is assumed that the tester will have a remote injector facility.
- Set the cable NVP according to the manufacturer's instructions.
- Understand whether it is the permanent link or the channel to be tested.
- Use the correct generic or proprietary tester interface as appropriate.
- Avoid manufacturer-specific test software wherever possible.
- Set the test machine to the correct standard, e.g. ISO 11801 2nd edition Class E permanent link.
- Ensure that the full suite of tests from IEC 61935 is being applied.
- Save the test results in the same format as the tester's own cable management software.

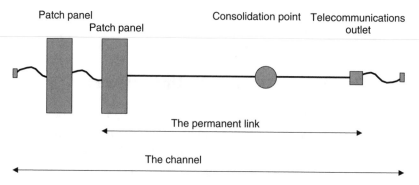

**Fig. 11.6** Permanent link and channel configurations.

- Ensure that every cable link has a single unique identifier.
- Do not mix copper and fibre test results in the same file.
- Replace test leads at the recommended interval.
- Use the latest and most appropriate software from the tester vendor.
- If a manufacturer's warranty is required, check what that manufacturer's testing requirements are.

Figure 11.6 is a reminder of the permanent link and channel definitions.

## 11.4   Optical fibre testing

The parameters of interest in an optical cable installation are:

- Link attenuation.
- Individual component attenuation.
- Return loss.
- System bandwidth.
- Link length.

Bandwidth is very difficult to test and no easily portable test equipment exists to do this. The designer and installer are in the hands of the optical cable supplier who will state the bandwidth of their product in terms of megahertz kilometres (MHz.km).

Return loss is the amount of light reflected back from connectors. It can be problematic for laser systems, especially analogue video systems where a high level of return loss close to the laser itself will either damage it or cause unacceptable signal distortion.

Link length relates to the overall link attenuation and propagation delay. Users are not really interested in the link length *per se* but the effects of the length, that is the attenuation and delay.

Individual component attenuation is of interest if a problem is suspected within a component such as a splice, a connector or a length of fibre itself, or if there is a contractual obligation to guarantee or prove the performance of a particular item.

For the short distances seen in structured cabling optical fibre links, the only test parameter of real interest is the overall link attenuation. This is the one figure that will give an unambiguous answer about whether the installer has done his or her job properly and give the strongest indication whether the optical link will actually work or not. The simplest and easiest way to test the link loss is with a power meter and stabilised optical source. Many people are familiar with the concept of an OTDR (optical time domain reflectometer), but as we shall see, the OTDR is a cumbersome and expensive way of conducting an acceptance test.

## 11.4.1   Power meter and light source testing

A stabilised light source is a laser or light-emitting diode that emits a steady stream of light, at a constant optical power, at one of the wavelengths normally used in optical fibre transmission. These are the first, second and third windows, which operate at around 850, 1300 and 1550 nm, respectively.

The stabilised light source injects its light into the optical link under test and at the receiving end is the power meter. This is a light sensitive diode device that will display a reading indicating how much power has been received through the optical link. The power meter will give the reading in decibels but for the figure to be valid the power meter has to know how much optical power was launched into the link. This is why calibration of the light source and the power meter is essential. Otherwise the power meter reading will be a purely random number. The calibration must take place before every sequence of tests and if the test lead from the light source is ever decoupled from the output connector, or if the light source is switched off, then the calibration must be done again.

The traditional method is to plug the light source into the power meter and turn it on. After a short period to allow the light source to settle down the power meter is 'zeroed'. This literally means setting the scale to read zero.

The test lead is then unplugged from the meter, never from the light source, and plugged into one end of the circuit under test. The power meter is then plugged into the receiving end of the link by another test lead. The total optical loss across the whole circuit, at the wavelength used is then read off the scale of the power meter. Remember that the light source is never turned off or disconnected from the test lead. Both test patch cords used must be of known good quality with scrupulously clean end faces. All that remains is to record the value and compare it with the optical loss calculated and expected for that class and length of link. If the measured reading is less than or equal to the calculated loss then the link has passed the acceptance test.

The standards only require a test from one end but many people prefer to test from both directions because a single ended test can mask some connector problems. A chipped connector, for example, can output light fairly well yet will reflect a large portion of light away when the circuit is reversed and it is attempted to launch light into it.

The link should also be tested at all the wavelengths at which it is expected to operate.

The test described above is relatively simple and straightforward but advances in connectors and test devices have tended to complicate matters. The first change is the advent of duplex optical connectors seen under the general banner of small form factor (SFF). The SFF style is typified by the MT-RJ, the LC, the Fiberjack and the VF-45(SG) connectors, amongst others. The philosophy of the small form factors connector is that it provides a duplex (i.e. two fibres) connection in the same footprint as a copper RJ-45 jack. This has the advantage of doubling up on the traditional port density of the older ST and SC connectors.

The disadvantage is that few power meters and light source combinations come with outputs that lend themselves to duplex connector operation and some imagination is required when it comes to organising the test and calibration leads.

Another change is the introduction of automated optical power meter 'add-ons' to existing copper cable testers. The advantages of these new test instruments are that the tester is the same, the process of testing is automated and the machine will give a simple pass/fail indication rather than just a number of decibels.

Unfortunately in attempting to give a simple pass or fail decision on the link, the tester has to make a number of assumptions about how many connectors there are in the link. This, coupled to confusion about calibrating with SFF connectors, has led many installers into a hopeless muddle when using these instruments.

The optical tester is a light source/power meter combination but it also acts as a simple OTDR to determine the length of the link. It simply 'pings' a pulse of light down one leg of the fibre, waits for the return and calculates the length by estimating the refractive index of the fibre and thus the velocity of the light in the fibre. The refractive index is the ratio of the speed of light in glass to the speed of light in a vacuum.

The tester measures the light power in the normal way, and knowing the length it can calculate the expected loss for the link and compare the two. It will also compare the result against the optical link tables from ISO 11801 2nd edition or one of the other standards. If the measured loss is less than the expectation from the standards, the machine can make its pass/fail decision. However, it is not link length that determines overall attenuation in a short fibre link, it is the performance of the connectors. It is imperative that the tester is told how many connectors and splices are in the channel it is measuring *and* it must be calibrated correctly, otherwise the results are meaningless. Installers must receive correct training to ensure they are following the set-up and operation instructions from the tester manufacturer. Many installers unfortunately adopt an 'open-the-box-and-just-plug-it-in' attitude to cable testing.

Testing and calibration is covered in the standards listed at the beginning of this chapter, e.g. IEC 61280 and EN 50346.

## *Power meter – light source: method 1*

This is described as Method 1 in EN 50346 and Method 1a in IEC 61280-4-2. The light source is plugged directly into the power meter via a test cord. After allowing the items to settle down the power reading is taken from the power meter. This value is recorded or the meter is zeroed. Test cord 1 (see Fig. 11.7) is then plugged into the link under test. Note the light source is not disconnected from the test cord or turned off at any stage during the test. Another test cord, test cord 2, is plugged in between the power meter and the other end of the link under test.

If the power meter has been zeroed, the link loss is simply read off the display. Otherwise the new power reading is subtracted from the original power reading to arrive at the same value. This method is the best when using individual fibre connectors that are hermaphroditic, that is there is no male or female section. Each connector is identical and mated together with an adapter such as in the ST or SC connectors.

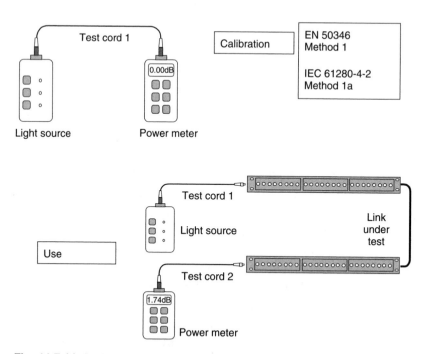

**Fig. 11.7** Method 1 light source/power meter test.

## Power meter – light source: method 2

Method 1 (IEC 61280 Method 1a) cannot cope with duplex connectors that have a specific male/female gender, for example the MT-RJ. MT-RJ patch panels are loaded with the male form of the MT-RJ, i.e. this is the half of the connector with the two pins. MT-RJ patch cords have the female form at either end. This prevents calibrating the power meter and light source by simply plugging one test lead between the two, even if the power meter is available with an MT-RJ output.

Method 2 therefore requires a third patch cord. This is demonstrated in Fig. 11.8. 'Method 2' is the EN 50346 terminology, IEC 61280-4-2 calls this 'Method 1c'. Method 1b requires plugging test cord 1 (called test 'jumper' in IEC 61280) directly into test cord 2 for the calibration exercise. There is potential confusion in terminology between EN 50346 and IEC 61280-4-2 because Method 2 in IEC 61280 means an OTDR test.

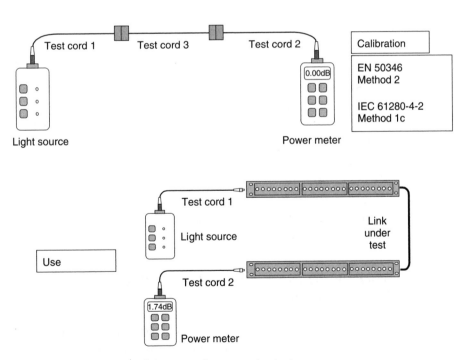

**Fig. 11.8** Method 2 light source/power meter test.

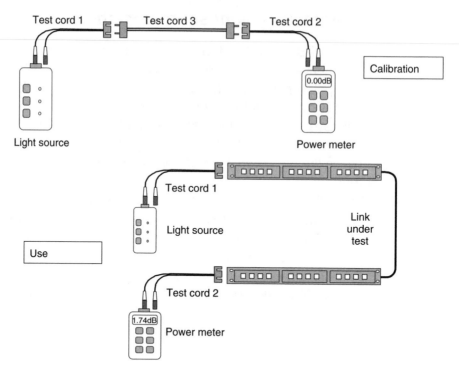

**Fig. 11.9** Method 2 light source power meter test with MT-RJ connectors.

Test cord 3 is not used in the test itself. For MT-RJ links a special duplex patch cord is required as test cord 3 where the MT-RJ connectors are of the male variety. Assuming that standard light sources and power meters are being used they will have a single SC or FC-PC connector fitted. Test cords 1 and 2 are therefore a female MT-RJ to two SC or FC-PC connectors. This method allows all the 'A' legs to be tested first in the system. Then the test instruments are recalibrated on the 'B' leg for all of the remaining links in the system. This is shown in Fig. 11.9.

## *Improving accuracy by reducing higher order modes*

When the light source launches light into the test cord, some of the energy will travel for a short while in the cladding and some will be

Table 11.3 Summary of mode stripping mandrel requirements

| Standards | Fibre type | | | Other requirements |
| --- | --- | --- | --- | --- |
| | 50/125 | 62.5/125 | Single mode | Cord length (m) |
| EN 50346 In turn quoting from EN 61300-3-34 | 5 turns around a 15-mm diameter | 5 turns around a 20-mm diameter | | 1–5 |
| IEC 14763-3 In turn quoting from IEC 61300-3-34 | 5 turns around a 15-mm diameter | 5 turns around a 20-mm diameter | | |
| TIA/EIA-568-B-1 | 5 turns around a 25-mm diameter for 0.9-mm tight buffered fibre. 5 turns around a 22-mm diameter for 3.0-mm tight buffered patch cord | 5 turns around a 20-mm diameter for 0.9-mm tight buffered fibre. 5 turns around a 17-mm diameter for 3.0-mm tight buffered patch cord | | |
| IEC 61280-4-2 | | | Two 80-mm diameter loops | 2–5 metres |
| IEC 61280-1-1 | | | One 80-mm diameter loop | 2–10 metres |

unevenly distributed across the fibre core in what is known as the higher order modes. After a few tens of metres of cable, the cladding modes and higher order modes have been stripped out and the power distribution is more even and predictable.

This phenomenon means that testing short links with relatively short test cords could lead to wide variations from one test to another. For such short link testing the standards recommend using a mode stripping mandrel or rod to improve accuracy and repeatability. This involves wrapping five non-overlapping turns around a mandrel of tight bend radius which forces all the cladding and higher order modes to escape from the fibre before they reach the connector. Table 11.3 summarises the requirements of various standards on mode stripping mandrels. The 22-mm and 17-mm mandrels appear to be most appropriate as most test cords are going to be 2.8–3.0-mm diameter patch cords. The single mode requirement of an 80-mm loop does not require a mandrel, just a loop.

## 11.4.2   Optical time domain reflectometer testing

An OTDR is a testing device that can identify faults within an optical cable and state accurately where they are. The OTDR can give a graphical characterisation of the fibre under test. The results, however, need expert interpretation and whereas the OTDR is an excellent fault finding device it is not best suited to simple accep-tance tests, especially for multimode fibre.

An OTDR works very much like radar. A short pulse of laser light is launched into the fibre and the light travels until it reaches the end of the fibre where some of it is reflected back to the source. This is known as Fresnel reflection which happens at the interface between materials of differing refractive index, such as air and glass. If we know the refractive index of the fibre under test, then we also know the speed at which light travels in that particular fibre. After this infor-mation is loaded into the OTDR it can time the point of departure of the light pulse until its return. With the round journey time recorded and knowledge of the refractive index (and hence velocity), the OTDR can work out accurately the overall length of the fibre to within a few metres.

Apart from this straightforward length measurement capability, the OTDR can also detect the small quantity of light continuously reflected back from small imperfections along the whole length of the fibre, known as Rayleigh scattering. The resulting trace will allow determination of the attenuation over any length of the fibre or indeed the whole cable run.

Any other discontinuity that does not break the fibre totally will also be seen, principally connectors, splices and tight-bend or compres-sion induced attenuation high spots. Figure 11.10 shows a typical OTDR trace.

The capabilities of an OTDR make it ideal for fault finding and characterising long distance telecommunication networks, but the

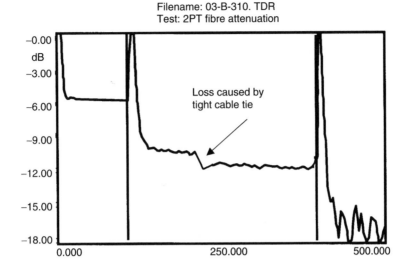

Fig. 11.10 Typical OTDR trace.

sophistication of the machine and the need to interpret results often lead to meaningless OTDR traces being presented to the end user in premises cabling projects. Figure 11.11(a), (b) and (c), shows three OTDR traces from three real installations. They are all fairly meaningless owing to the very short lengths of fibre attempting to be tested, lack of launch and tail leads and excessive pulse lengths used.

IEC 61280-4-2 calls OTDR tests Method 2. A power meter test is Method 1. If the results differ between tests done with Methods 1 and 2 then the result from Method 1, i.e. the power meter test, shall prevail, according to IEC 61280. IEC 61280-4-2 does not call for a tail lead, only a launch lead, which it calls a 'dead-zone fibre'; however, the standard does require testing from both directions and the results averaged to give an accurate figure.

Link loss can be determined with an OTDR but it must be done by placing the cursors over the link element of interest and deducing the loss between the cursors. This should be done from both directions and averaged if an accurate result is required.

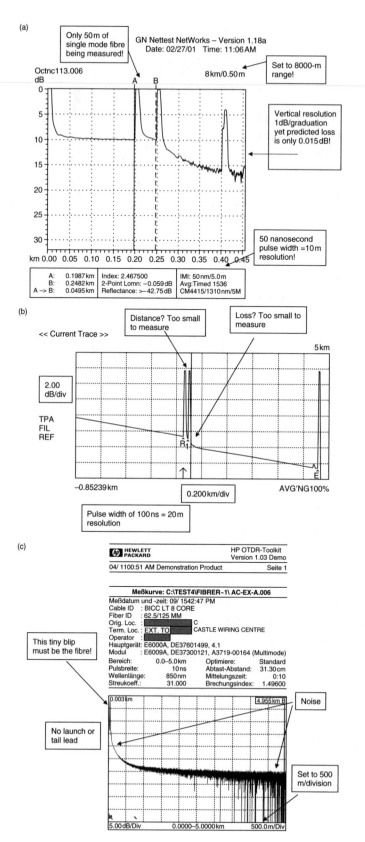

**Fig. 11.11** (a), (b) and (c) Unusable OTDR traces from a real installation.

## OTDR problems

- *Pulse length and dynamic range.* To see detail over a 50-km cable length a powerful laser pulse has to be sent. This equates with the need for a long pulse duration. The length of the pulse of light is equivalent to tens of metres of optical fibre. This gives a low resolution, as events closer to each other than the pulse width will not be seen. There will also be a dead zone in front of the launch device, as the receiver will be closed off until the light pulse has finished.

- *Dead zone 'blinding'.* The amount of light power reflected back from distant events is very small. Local events such as a connector pair a few tens of metres away will reflect back a proportionally much larger signal. This signal can temporarily blind the receiver and it takes a short time to recover. But this recovery period equates with a dead zone forming behind the local event that cannot be seen.

- *Ghosting.* If a lot of power is reflected back from a close event such as a connector pair then some of it may reflect back again from the OTDR towards the connector pair. This can happen several times. The multiple reflected signal will take time to make the round journey and so the OTDR will interpret them as other attenuation events happening at exact multiples of the original event distance.

- *Fibre numerical aperture mismatch.* A splice or a connection onto a fibre with a different numerical aperture can reflect back more light than the original part of the fibre. The OTDR will interpret this as a gain in the fibre link rather than as an attenuation. The loss of the fibre link can only be accurately determined by measuring it from both directions and averaging the two results.

- *Insufficient averaging.* To see the returning signal the fibre is scanned several times and the results averaged out. This has the effect of cancelling out much of the noise because the noise is random in nature whereas the desired signal will remain the same through every scan. If the OTDR is not allowed enough time to settle down and average the result then the trace will be too mixed in with noise to be readable.

The operator of the OTDR must therefore do the following to obtain meaningful results:

- Load the correct value for the refractive index.
- Set up an appropriate pulse width for the length of fibre being tested.
- Use long enough launch and tail leads, that is, extra fibre added at the beginning and end of the real cable link, so that dead zones at both ends of the fibre are removed and there is sufficient resolution to see the far end connector. Figure 11.12 shows the launch–tail lead configuration.
- Allow sufficient time for the OTDR to average out the noise and signal.
- Test from both ends.
- Test at all wavelengths likely to be used, e.g. 850 and 1300nm.
- Use the OTDR cursors so that the length of fibre under investigation is measured accurately.
- Set up the OTDR and use it according to the manufacturer's instructions. Typical refractive index figures are given in Table 11.4.

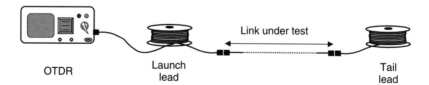

**Fig. 11.12** Launch and tail lead on an OTDR test.

| Table 11.4 Typical refractive indices of optical fibres | | | |
| --- | --- | --- | --- |
| | 850nm | 1300nm | 1310nm |
| 50/125 | 1.481 | 1.476 | |
| 62.5/125 | 1.495 | 1.490 | |
| Single mode | | | 1.472 |

## 11.4.3    Recording the optical test results

Hard copies of the OTDR traces must be supplied as part of the final documentation. It is a good idea for somebody who understands OTDR traces to have a look at them and to accept them formally. Many OTDRs store the information in electronic format but in a file format unique to the OTDR manufacturer. This makes it difficult or impossible for anybody to open them again in the future.

Power meter testing should be done on all links, at 850 and 1300 nm for multimode and at 1310 nm and maybe 1550 nm for the single mode fibres, in both directions. The results must be presented on a test sheet as shown in Fig. 11.13. Note in this chart that there is a column for calculating the expected attenuation values according to ISO 11801 rules. The four columns on the right are the actual measurements taken at both wavelengths (multimode) and in both directions. If all of the measured values are equal to or less than the calculated values, then the system has passed the test and can be accepted. Another common fault of optical cable testing is simply to record lists of attenuation readings. Without referring to the attenuation that is expected for that link, a simple list of attenuation values is meaningless.

The power meter and light source must be correctly calibrated according to the manufacturer's instructions. The recent addition of optical power meter add-ons to hand-held copper testers has been a mixed blessing. A simple power meter/light source combination will give a simple value for attenuation and it is up to the installer to decide if that value is acceptable. A fibre optic power meter 'add-on', however, wants to make a pass/fail decision, as it does for the copper tests. It is essential therefore that fibre tester 'add-ons' are calibrated exactly according to their manufacturer's instructions. Unfortunately, there have already been many projects where large test regimes have been completely wasted owing to reliance on meaningless test reports from these types of testers. The add-on power meter will save the results electronically, and as already stated for the copper results from the same tester, the file format must be the same as the cable management software from that same tester manufacturer. Do not mix up copper and fibre tests in the same file!

| OPTICAL TEST REPORT | SHEET REF. NO. .............. DATE..................... |
|---|---|

INSTALLATION Co.......................................................................................
END-USER          .....................................................................................
SITE ADDRESS      .....................................................................................
                  .....................................................................................

CABLE I/D: ..........................................
TUBE COLOUR OR NUMBER FROM REFERENCE TUBE: ..........................

| CABLE LENGTH: ............................ | No. OF FIBRES IN TUBE............. |
|---|---|
| FIBRE TYPE: ................................. | No. OF FIBRES IN CABLE: ......... |

CABLE TYPE/PART NO: ..............................................................................
PANEL TYPE/PART NO: ..............................................................................
CONNECTOR TYPE/PART NO: ....................................................................

| No. OF SPLICES IN LINK .............. | ROUTE, FROM ....................... |
|---|---|
| No. OF CONNECTORS IN LINK ...... | TO ......................................... |

WAVELENGTHS TESTED  ...........................
MAKE & MODEL OF TEST EQUIPMENT  ......................................................

| FIBRE NO. | FIBRE COLOUR | CALCULATED ATTENUATION | | MEASURED ATTENUATION | | | |
|---|---|---|---|---|---|---|---|
| | | 850 nm | 1300 nm | A to B | | B to A | |
| | | | | 850 nm | 1300 nm | 850 nm | 1300nm |
| 1 | | | | | | | |
| 2 | | | | | | | |
| 3 | | | | | | | |
| 4 | | | | | | | |
| 5 | | | | | | | |
| 6 | | | | | | | |
| 7 | | | | | | | |
| 8 | | | | | | | |
| 9 | | | | | | | |
| 10 | | | | | | | |
| 11 | | | | | | | |
| 12 | | | | | | | |

TESTER NAME.................….   SIGNATURE...............   DATE.........

**Fig. 11.13** Optical test report.

## 11.4.4 Determining the pass or fail parameters for optical fibre testing

Having ascertained the link loss, we have to decide if it is acceptable under the rules of ISO 11801 2nd edition. To do this the measured results are compared against those shown in Table 11.5. We can see that the links are specified as 300-m, 500-m or 2000-m links. For each operating wavelength there is a maximum expected attenuation for each class of link. Table 11.6 gives the allowance for the optical loss for each item encountered in the optical cabling system, that is, the connector pairs, optical splices and the fibres themselves. To arrive at the predicted overall link loss it is just a matter of adding up

Table 11.5 ISO 11801 2nd Edition, channel attentuation allowance

| Cabling subsystem | Link length (m max) | Channel attenuation (dB max) | | | |
| --- | --- | --- | --- | --- | --- |
| | | Multimode | | Single mode | |
| | | 850 nm | 1300 nm | 1310 nm | 1550 nm |
| OF-300 | 300 | 2.55 | 1.95 | 1.8 | 1.8 |
| OF-500 | 500 | 3.25 | 2.25 | 2.0 | 2.0 |
| OF-2000 | 2000 | 8.5 | 4.5 | 3.5 | 3.5 |

Table 11.6 ISO 11801 2nd Edition optical component loss allowance

| Component | 850 nm multimode | 1300 nm multimode | 1310 nm single mode and 1550 nm single mode | | |
| --- | --- | --- | --- | --- | --- |
| | | | Inside plant | Outside plant | |
| Optical cable | 3.5 dB/km | 1.5 dB/km | 1.0 dB/km | TIA 0.5 dB/km | ISO 1.0 dB/km |
| Optical connectors | 0.75 dB | 0.75 dB | 0.75 dB | 0.75 dB | 0.75 dB |
| Optical splice | 0.3 dB | 0.3 dB | 0.3 dB | 0.3 dB | 0.3 dB |

Note: all cable measurements are dB/km.
All figures are the same for ISO 11801 edition and TIA/EIA 568B except where identified by 'TIA' or 'ISO'.

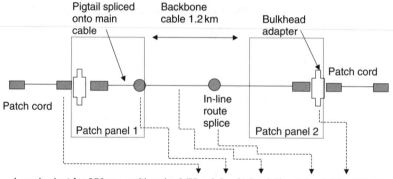

Loss budget for 850 nm multimode, 0.75 + 0.3 + (1.2 x 3.5) + 0.3 + 0.75 = 6.3 dB

Loss budget for 1300 nm multimode, 0.75 + 0.3 + (1.2 x 1.5) + 0.3 + 0.75 = 3.9 dB

Loss budget for 1310 nm single mode, 0.75 + 0.3 + (1.2 x 1.0) + 0.3 + 0.75 = 3.3 dB

**Fig. 11.14** Calculating optical link loss.

each individual loss. This is the beauty of working in decibels. Figure 11.14 shows an example of adding up all the individual component losses for each item in the cable link. Notice that for fibre the attenuation is different for each wavelength, so the calculation should be done for both wavelengths. To arrive at the fibre link loss, simply multiply the manufacturer's published fibre attenuation specification (or use the standard ISO 11801 2nd edition figures) by the length, in kilometres. For example, if the fibre is rated at 3.0 dB/km at 850 nm and the link is 500 m long (half a kilometre) then the link fibre attenuation is $0.5 \times 3.0 = 1.5$ dB.

To this figure we add the loss for each component, for example, if there are two connectors (0.75 dB each) and one splice (0.3 dB each) then we have $0.75 + 0.75 + 0.3 = 1.8$ dB plus the fibre loss of 1.5 dB, giving a total of 3.3 dB. We would arrive at a different value if we did the calculation at 1300 nm.

The optical design and test methodology is therefore simple:

• Design the system to work according to ISO 11801 2nd edition rules.
• Install the optical system.
• Test the link loss for each class or section.
• Determine if the measured link loss is less than the expected loss.

- If the measured link loss is lower, then the system has passed.
- If the measured link loss is greater, then the link has failed.

If the link has failed the attenuation test then corrective action must be taken. Connectors are the first things to look at, as they are the most likely source of problems. After that, splices, and after that the cable run itself. Fault finding is where an OTDR becomes invaluable. A power meter/light source test will determine if there is a problem or not but it will not determine where or what that problem is. An OTDR can pinpoint the exact cause of the problem.

## 11.4.5   Air blown fibre testing

Air blown fibre consists of empty plastic tubes, generally 5–8 mm in diameter, into which bundles of fibres or individual optical fibres are blown using compressed air, or other gases, at a later date. Once the optical fibres have been installed they are terminated and tested in exactly the same manner as any other kind of optical fibre. The blown-fibre ducts, however, need a particular kind of test after installation to ensure they have not been damaged, which would of course preclude any fibre blowing in the future.

The blown-fibre duct test is in two parts, but it is conducted by the same machine or device. A device injects a steel ballbearing into one end of the duct under pressure from compressed air. If the ballbearing reaches the special valve at the far end, the installer knows that the duct has not been crushed or kinked. The specially designed valve at the far end is sealed by the arrival of the ballbearing and the pressure in the duct will start to rise if the source of compressed air is still attached. The duct is allowed to rise to its normal maximum working pressure, typically ten times atmospheric pressure, and if the duct maintains that pressure over a few minutes, it confirms that the duct has not been punctured or ripped anywhere. The duct is then sealed at both ends to prevent the ingress of water, dust and insects, and it should be possible to return to that duct at any time in the future to blow fibre. If a long time has elapsed, however, then it would be wise to pressure test the duct again to ensure that no physical damage has occurred to the ducts in the intervening years. If the ducts are damaged, the optical fibre will not blow into the duct or it may enter the duct only to stop at the point of damage. The fibre consumed would normally be lost.

Very short runs, like those that may be encountered in FTTD may not warrant the added cost of pressure tests as the chances of problems are very slim.

# 12

# Step-by-step guide to designing a structured cabling system

This chapter is intended as a summary of the processes dealt with in previous chapters to provide a step-by-step guide to designing a structured cabling system. See individual chapters for more detail.

## 12.1 Design checklist

- First pick a standards philosophy.
- Decide upon the overall topology and density of outlets.
- Decide upon category of cable and class of optical fibre.
- Decide upon unscreened or screened copper cable.
- Decide upon fire rating of internal cables.
- Decide upon type of optical connector.
- Locate and specify the spaces.
- Determine the nature of the containment system.
- Determine earthing and bonding requirements.
- Decide upon the administration system.
- Identify the civil engineering/rights of way issues for external cabling.
- Identify specific installation instructions.
- Decide upon the testing and handover regime.
- Prequalification and bidding.

## 12.2   First pick a standards philosophy

Step 1 in designing or procuring a structured cabling system is to decide which standards philosophy to adhere to. The choices are:

- None at all, just specify a grade of cabling and let the installer sort it all out.
- Pick a famous brand and hope that the ministrations of a large corporation will ensure that something relevant, useful and reliable is installed.
- Pick and mix from different standards.
- Use local/national standards.
- Use international standards.

Leaving the choice of components and the resulting performance totally in the hands of an installer is a risky business and should only be contemplated by those looking for a very undemanding network at the absolute lowest price. However, do not forget the ongoing cost of ownership when things start to go wrong. Many surveys point out that more than half of all network faults are cable related.

The days of picking and specifying a major brand name as the sole supplier are probably over with the advent of comprehensive international standards, although it is through the efforts of some of the major suppliers that serious standards and well-engineered products do exist. Many of the larger users now tend to 'prequalify' a group of major manufacturers' products, after specifying them according to the standards, and then prequalifying a group of competent installers, telling them they can bid any product from the pre-approved list.

The pick-and-mix approach to standards is not necessary either, as full families of standards now exist across all the principal standards-writing bodies. A strange mix of standards usually says more about the specification compiler's lack of knowledge rather than anything else. However, nobody has a standard for everything and an open mind is needed to get the best overall package put together to address all network problems. Remember nowadays that the cabling installation does not just require an overall design standard but needs information relating to EMC/EMI, fire performance, cable containment, earthing and bonding and a host of other related subjects.

Table 12.1  Cabling design standards for different regions

| European Union | United States | Canada | Australia and, New Zealand | Rest of the World |
|---|---|---|---|---|
| EN 50173 | TIA/EIA 568 | CAN/CSA-T529 | AS/NZS 3080 | ISO 11801 |

There are few local standards relating to cabling, apart from some fire regulations, but there are national standards for the USA, for Canada, for Australia and New Zealand and also for the European Union.

Systems located in the USA should design cable standards to the ANSI/TIA/EIA range of standards with their supporting National Electrical Codes. The principal design standard in the USA is TIA/EIA 568B. The equivalent standard for Canada is CAN/CSA T529 and in Australia and New Zealand it is AS/NZS 3080.

In the European Union (EU) the relevant standards are written by CENELEC, and the standard for structured cabling is EN 50173. All CENELEC standards start with EN for European Norm.

World standards are written by ISO and IEC. Both are based in Geneva, Switzerland, but they are truly international bodies with representatives from all continents forming their committee membership. Their standards are written for a worldwide audience and can be invoked in the EU and the USA with full impunity. In the EU it is expected that CENELEC standards will be utilised, where they exist, when users are spending public money. The EU Procurement Directive looks to European standards to ensure fair and open competition across all the member states.

Table 12.1 summarises the overall standards options.

See Chapter 2 for more details.

## 12.2.1  Procurement guide

The following form of words is offered as a suggestion when specifying the standards compliance of the prospective cable system:

The structured cabling system must conform to the overall technical requirements specified in the following principal standards including all referenced and subsidiary standards invoked by the principal standard except where more completely described in this specification. The contractor must identify where their offer deviates from the principal standard or any relevant referenced or subsidiary standards.

And then pick one from the following principal standards:

- ISO 11801 2nd edition: *Information technology – Cabling for customer premises*.
- EN 50173 2nd edition: *Information technology – Generic cabling systems*.
- ANSI/TIA/EIA-568-B: *Commercial Building Telecommunications Cabling Standard*.

# 12.3   Cable system topology and density

The cable system must conform to one of the ISO 11801 standard models as described in Chapter 3. The user does, however, have a number of choices.

## 12.3.1   How many layers in the hierarchy are required?

There can be up to three layers in the hierarchical structure of ISO 11801 2nd edition, EN 50173 and TIA/EIA-568-B:

- Horizontal cabling.
- Building backbone cabling.
- Campus backbone cabling.

## 12.3.2   Is centralised optical architecture required?

Centralised optical architecture (COA) or FTTD may be appropriate, in place of copper, where the following criteria are met:

- Average cable runs from the TO to the telecommunications room or the equipment room are in excess of 90 m.

- Electromagnetic interference problems are likely to be severe, e.g. airports, power stations, etc, with interfering fields regularly in excess of 10 V/m.
- The need for security is paramount, e.g. military and banking installations.

The user is urged to look at the whole costs of a project when considering COA or FTTD and not just individual component costs. The cost of the transmission equipment represents a large portion of the overall cost and the difference between selecting, for example, 100BASE-SX over 100BASE-F equipment will make a significant impact on the project costs. There are also other hidden 'savings' when using COA. These arise from the management overheads to be saved (including cost of floor space, air conditioning, IT department allocation, etc.) when doing away with active equipment in the telecommunications room or doing away with the telecommunications room altogether.

## 12.3.3  Horizontal cabling model choices

The user has a number of choices in the topography of the horizontal cable. These are all shown pictorially in Chapter 3.

- *Interconnect model (2-connector)*. This is the simplest and cheapest form consisting of the TO and one patch panel connected by up to 90 m of the horizontal cable.
- *Cross-connect model (3-connector)*. With cross-connect two patch panels are needed or alternatively two rows of IDC blocks. One set of patch panels is dedicated to the transmission equipment and the other is dedicated to the horizontal cabling. A jumper connects the two panels. Cross-connect gives the greatest flexibility when it comes to patching users to applications. In very large installations it can be very difficult to get all of the patch panels physically close enough to all of the LAN equipment if only using the interconnect model. The extra flexibility comes at the price of buying twice as many patch panels, twice the number of patch cords and occupying twice the rack space. Also there will be twice as much crosstalk introduced into the cabling sys-

tem but modern cabling systems and components should be designed to cope with up to four connectors in the horizontal channel. Cross-connect is also referred to as *double representation*. With the current state of technology cross-connect is necessary to implement intelligent patching.

- *Consolidation point.* A CP is an extra joint added into the horizontal cabling to improve flexibility. It is most often seen in large open-plan office designs where the basic horizontal cabling is done early in the project with the final connections to the desk made later.
- *All possible combinations.* We can see from the above and Chapter 3 that all the possible combinations are:
  - Interconnect and TO – two connector model.
  - Interconnect and CP and TO – three connector model.
  - Cross-connect and TO – three connector model.
  - Cross-connect and CP and TO – four connector model.

### 12.3.4   Using a multi-user telecommunications outlet

A MUTO is a multi-way outlet, usually up to twelve. It was formally known as a MUTOA (multi-user telecommunications outlet assembly). The MUTO can be used to give longer lengths of work area cable at the expense of shortening the fixed horizontal cabling. This is not a one-to-one swap, as the attenuation of the work area patch cord is usually higher than the horizontal solid-core cable. It is not necessary to use a MUTO to achieve the longer work area cable length as the rules in Table 12.2 will apply to any TO assembly.

Table 12.2 Using extended length work area cables/patch cords

| Cabling model | Category 5 cable Class D | Category 6 cable Class E |
|---|---|---|
| Interconnect–TO | H=109-FX | H=107-3-FX |
| Cross-connect–TO | H=107-FX | H=106-3-FX |
| Interconnect–CP–TO | H=107-FX-CY | H=106-3-FX-CY |
| Cross-connect–CP–TO | H=105-FX-CY | H=105-3-FX-CY |

H = max length of horizontal cable (metres); F = combined length of all patch cords; C = length of CP cable; X = ratio of flexible cable attenuation to horizontal cable attenuation; Y = ratio of CP cable attenuation to horizontal cable attenuation.
[1] If X and Y are not known then use a factor of 1.2 for all UTP patch cord and 1.5 for all screened patch cord
[2] Although ISO 11801 second edition allows you to calculate channel lengths when using reduced length horizontal cable and longer patch cords, for Class D and E operation it is still a maximum of 90 m for the fixed horizontal cabling and 100 m for the overall channel.

## 12.3.5   Cable lengths

The maximum permissible cable lengths are defined in Chapter 3. The most important one is the 90-m limit on the horizontal cabling. If the desired system design cannot be achieved within this constraint then the options are:

- Redesign the cabling layout so that the telecommunications room is closer to the TOs.
- Use optical fibre.
- Downgrade the class of operation expected, e.g. Category 5,6 or 7 cable will be downgraded to Class C, B or A system operation depending upon the distance covered. Table 12.3 gives the rules to be followed in these circumstances.

So, for example, a Category 5 cable running a distance of 120 m would be described as Class C operation and the system frequency range expected would be 16 MHz. We must recognise that extending the length of the horizontal cabling by derating the achievable class of operation is not strictly in the spirit of ISO 11801 2nd edition.

Table 12.3 Derating class of operation when using extended lengths

| Class of application | Class A | Class B | Class C | Class D | Class E |
|---|---|---|---|---|---|
| Frequency range | 100 kHz | 1 MHz | 16 MHz | 100 MHz | 250 MHz |
| Max distance with **Cat 3*** cable | 2000 m | 200 m | 90 m** | — | — |
| Max distance with **Cat 5** cable | 2000 m | 250-FX | 170-FX | 105-FX** | — |
| Max distance with **Cat 6** cable | 2000 m | 260-FX | 185-FX | 111-FX | 102-FX** |

[1]F = combined length of patchcords/jumpers in the channel; [2]X = ratio of extra loss due to flexible cable used in patch cords. If this figure is not known then use factors of 1.2 for UTP cords and 1.5 for screened cords as a default; [3]* Cat 3 components are no longer supported by ISO 11801 second edition. They may still be referenced though from ISO 11801 Edition 1.2 (2000); [4]Cat 3 cabling should be used in the *backbone* only; [5]** Maximum channel length is 100 m for these categories of cable and classes of operation; [6]The Cat 5 and 6 figures above are taken from ISO 11810 second edition. *Backbone cabling model*, and this edition only recognises horizontal cabling up to a maximum of 100 m. Check that the manufacturer supports longer distances, when using Cat 5 and Cat 6 cabling, in the horizontal as well.

The Category 5 and 6 values in Table 12.3 are taken from the ISO 11801 2nd edition *Backbone cabling model*, and this edition only recognises horizontal cabling up to a maximum of 100 m. Table 12.3 is offered as a potential strategy when 90 m is not achievable without great expense. It would be wise to check that the cable manufacturer or supplier also supports this interpretation if a manufacturer's warranty is expected or required.

## 12.3.6   Density and location of TOs

There is no single recommendation for the distribution and density of TOs within premises. However, each individual work area must be served by a minimum of two TOs. The first outlet must be a four-pair balanced cable of Category 5, 6 or 7. The second outlet may be the same or a pair of optical fibres.

The location of the TOs should minimise the length requirement of the work area cable needed to connect the user's devices to the TO. This would normally be within 3 m but up to 20 m is possible. However, for normal office environments the 3-m distance is the desired target. In a normal office environment, a density of around two outlets per 10 m² of useable floor space would be expected. Dealer desk environments could have a density up to six times larger than this.

Thought also needs to be given to the physical presentation of the TOs. There are usually three choices:

- Wall mounted, dado rail height.
- Wall mounted, skirting board height.
- Floor mounted.

Wall mounted outlets require details of the type and density, for example, British Standard 86 × 86 mm single gang dual outlet would be one style to specify. The UK and Ireland tend always to use single or double gang BS wall outlets for all electrical and communications wall outlets. Other countries tend to have a wider variation of styles in circulation. Local expectations, style and practice should be taken into account when designing wall outlets.

One final detail is the issue of shutters over the face of the outlet. In the UK there is an expectation that there should be shutters for all wall-mounted outlets and a belief that this is required by regulation. This is not the case as there is no such regulation and the user is free to choose whether he/she wants shutters or not.

The details of the TO presentation and trunking arrangement should be considered as part of the cable containment strategy.

# 12.4   Picking the category of cable and class of optical fibre

Cables can appear in three locations, the horizontal, building backbone and campus backbone. The cable requirements must be defined for all three where they exist. Chapter 4 describes the products in more detail and Chapter 5 addresses the issues of picking cable media according to future bandwidth requirements.

## 12.4.1   Copper cable

- *Identify location*. Horizontal, building backbone or campus backbone.
- *Select category, 3, 5, 6 or 7*. Category 3 is for backbones only.
- *Specify characteristic impedance*. Select 100 Ω unless specific reasons require otherwise (Chapter 4).
- *Select cable construction*. Four-pair units are the most common but other pair counts are possible; or else bundled constructions such as Siamese (two four-pair cables), bundles of four-pair units or hybrids of copper and optical cables or even air-blown-fibre ducts.
- *Select unscreened or screened*. If the cable is to be screened then there is a choice of foil screen (FTP) or foil and braid screen (S-FTP) or individual pair screen and overall screen (STP/PIMF, pairs in metal foil). See Chapter 6.
- *Select fire performance for internal cables*. In the USA select general purpose, riser or plenum grade according to location. In Europe select the appropriate Euroclass or IEC rating. For the rest of the world select appropriate IEC rating unless local more stringent conditions apply. Note that some short lengths may be run outdoors for which there are various types of outdoor grade 'horizontal' cables available from different manufacturers. Chapter 7 gives full details of fire performance options available.
- *Other cable parameters*.
  - *Colour*: Most manufacturers produce their indoor copper cables in grey and this is likely to be the most cost effective route to pursue. Any colour can be made, of course, but this will usually mean a higher price, longer lead time and larger minimum order quantities.
  - *Other information*: Installers may wish to know what markings are on the cable such as 'metre marking'. This means printing an ascending number along the cable sheath at 1-m intervals (albeit with no more than ±2% accuracy) so that the installer

can tell how long the cable route is and how much is left in the box or on the reel. Some manufacturers print the cable nominal velocity of propagation (NVP) on the sheath to assist installers in putting the right value into the hand-held tester. If the cable is recognised, approved or listed by a third party test house, this will probably also appear on the sheath. This might be UL or ETL in the USA or Europacable in Europe. Installers may also wish to specify if the cable is to come in a box, usually 305 m (1000 feet) or on a drum/reel that will usually be 500 m.

## 12.4.2   Procurement guide – copper cable

The following standards can be used when specifying the cable system.

### *International*

- IEC 61156: *Multicore and symmetrical pair/quad cables for digital communications.*
- IEC 61156-5: *Symmetrical pair/quad cables with transmission characteristics up to 600 MHz.*
- IEC 61156-6: *Symmetrical pair/quad cables with transmission characteristics up to 600 MHz – work area wiring.*

### *Europe*

- EN 50288: *Multi-element metallic cables used in analogue and digital communications and control. (Note: replaces EN 50167 EN 50168 and EN 50169)*
- EN 50288-2-1: *100 MHz screened, horizontal and backbone.*
- EN 50288-2-2: *100 MHz screened, patch.*
- EN 50288-3-1: *100 MHz unscreened, horizontal and backbone.*
- EN 50288-3-2: *100 MHz unscreened, patch.*
- EN 50288-4-1: *600 MHz screened, horizontal and backbone.*
- EN 50288-4-2: *600 MHz screened, patch.*

- EN 50288-5-1: *200 MHz screened, horizontal and backbone.*
- EN 50288-5-2: *200 MHz screened, patch.*
- EN 50288-6-1: *200 MHz unscreened, horizontal and backbone.*
- EN 50288-6-2: *200 MHz unscreened, patch.*

## *America*

The horizontal copper cables are described in TIA/EIA-568-B.2, but more details can be found in the following US standards:

- NEMA WC-63.1: *Performance standards for twisted pair premise voice and data communications cable.*
- NEMA WC-66.1: *Performance standards for Cat 6, Cat 7 100-Ω shielded and unshielded twisted pair cables.*
- ICEA S-80-576: *Communications wire and cable for wiring of premises.*
- ICEA S-90-661: *Individually unshielded twisted pair indoor cables.*
- ICEA S-100-685: *Station wire for indoor/outdoor use.*
- ICEA S-101-699: *Cat 3 station wire and inside wiring cables up to 600 pairs.*
- ICEA S-102-700: *Cat 5, 4-pair, indoor UTP wiring standard.*
- ICEA S-103-701: *AR&M riser cable.*

### 12.4.3  Optical cable and fibre

## *Select the optical fibre*

Table 12.4 is a reminder of the speed versus length criteria, see also Chapter 4. Although other fibre grades are available, the ISO 11801 2nd edition choices are as follows:

- OM1  (50/125 or 62.5/125).
- OM2  (50/125 or 62.5/125).
- OM3  50/125.
- OS1  single mode.

Tables 12.5, 12.6 and 12.7 give more design detail to address the requirements of individual named protocols.

Table 12.4 Optical fibre specification by distance and speed

| Speed (Mb/s) | Distance (m) | | |
|---|---|---|---|
| | 300 | 500 | 2000 |
| 10 | OM1 | OM1 | OM1 |
| 100 | OM1 | OM1 | OM1 |
| 1000 | OM1 | OM2 | OS1 |
| 10 000 | OM3 | OS1 | OS1 |

Table 12.5 Optical LAN operation by power budget

| Network application | Optical power budget | | | | | |
|---|---|---|---|---|---|---|
| | Multi mode | | | | Single mode | |
| | 850 nm | | 1300 nm | | 1310 nm | |
| 10BASE-FL,FB | 12.5 (6.8) | 12.5 (7.8) | | | | |
| 100BASE-FX | | | 11.0 (6.0) | 11.0 (6.3) | | |
| 1000BASE-SX | 2.6 | 3.2 (3.9) | | | | |
| 1000BASE-LX | | | 2.35 | 4.0 (3.5) | 5.0 | 4.7 |
| Token ring 4,16 | 13.0 (8.0) | 13.0 (8.3) | | | | |
| ATM 155 | | 7.2 | | 10.0 (5.3) | 7.0 | 7.0 to 12.0 |
| ATM 622 | | 4.0 | | 6.0 (2.0) | 7.0 | 7.0 to 12.0 |
| Fibre channel 1062 | | 4.0 | | | 6.0 | 6.0 to 14.0 |
| FDDI | | | 11.0 (6.0) | 11.0 (6.3) | 10.0 | |

The figures in parentheses are for 50/125 performance.
ISO 11801 2nd edition in normal type, *TIA/EIA-568-B.1 in italics*.

Table 12.6 Optical LAN operation by maximum supportable distance

| Network application | Maximum supportable distance | | | | | |
|---|---|---|---|---|---|---|
| | Multi mode | | | | Single mode | |
| | 850 nm | | 1300 nm | | 1310 nm | |
| 10BASE-FL,FB | 2000 (1514) | *2000 (2000)* | | | | |
| 100BASE-FX | | | 2000 (2000) | *2000 (2000)* | | |
| 1000BASE-SX | 275 (550) | *220 (550)* | | | | |
| 1000BASE-LX | | | 550 550 | *550 (550)* | 2000 | *5000* |
| 100BASE-SX* | 300 (300) | | | | | |
| Token ring 4,16 | 2000 (1571) | *2000 (2000)* | | | | |
| ATM 155 | 1000 (1000) | *1000 (1000)* | 2000 (2000) | *2000 (2000)* | 2000 | *15000* |
| ATM 622 | 300 (300) | *300 (300)* | 500 (330) | *500 (500)* | 2000 | *15000* |
| Fibre channel 1062 | 300 (500) | *300 (500)* | | | 2000 | *10000* |
| FDDI | | | 2000 (2000) | *2000 (2000)* | 2000 | *40000* |

The figures in parentheses are for 50/125 performance, otherwise 62.5/125 fibre.

ISO 11801 2nd edition in normal type, *TIA/EIA-568-B.1 in italics*.

*100BASE-SX is described in TIA/EIA-785.

Table 12.7 ISO 11801 2nd edition optical channel support summary

| Network application | ISO 11801 Channel support (summary only) | | | | | | |
| --- | --- | --- | --- | --- | --- | --- | --- |
| | OM1 | | OM2 | | OM3 | | OS1 |
| | 850nm | 1300nm | 850nm | 1300nm | 850nm | 1300nm | 1310nm |
| 10BASE-FL,FB | OF 2000 | | OF 2000 | | OF 2000 | | |
| 100BASE-FX | | OF 2000 | | OF 2000 | | OF 2000 | |
| 100BASE-SX** | OF 300 | | OF 300 | | OF 300 | | |
| 1000BASE-SX | OF* 300 | | OF 500 | | OF 500 | | |
| 1000BASE-LX | | OF 500 | | OF 500 | | OF 500 | OF 2000 |
| 10GBASE-LX4/LW4 | | OF 300 | | OF 300 | | nys*** | OF 2000 |
| 10GBASE-ER/EW | | | | | | | OF 2000 (1550nm) |
| 10GBASE-SR/SW | | | | | OF 300 | | |
| 10GBASE-LR/LW | | | | | | | OF 2000 |
| Token ring 4,16 | OF 2000 | | OF 2000 | | OF 2000 | | |
| ATM 155 | OF 500 | OF 2000 | OF 500 | OF 2000 | OF 500 | OF 2000 | OF 2000 |
| ATM 622 | OF 500 | OF 500 | OF 500 | OF 500 | OF 500 | OF 500 | OF 2000 |
| Fibre channel 1062 | OF 500 | | OF 500 | | OF 500 | | OF 2000 |
| FDDI | | OF 2000 | | OF 2000 | | OF 2000 | OF 2000 smf-pmd |

*Note that IEEE 802.3z quotes 220 m for 160 MHz.km fibre and 275 m for 200 MHz.km fibre.
**100BASE-SX is described in TIA/EIA-785.
***nys: not yet specified.

## Fibre delivery method

Select from the following:

* Optical cable.
* Hybrid optical/copper cable.
* Air blown fibre.

## How many fibres are required?

For FTTD applications there will be at least two optical fibres per user but it is possible to run an eight-fibre cable, for example, into a zone expected to service four users, or to run an air blown fibre duct capable of carrying up to 12 fibres. Backbone cables will normally have at least eight fibres. See Chapter 5 for more advice on selecting the appropriate number of optical fibres.

## Optical cable design

The optical cable design must be robust enough for the environment in which it will be installed. It is usually sufficient for an installer and manufacturer to agree on the physical parameters but standards, such as IEC 60794, can assist. An optical cable is specified along the following lines:

* Number of fibres.
* Type of fibres.
* Environment of use:
    - indoor grade.
    - outdoor duct grade.
    - outdoor armoured (direct burial).
    - aerial.

Note that internal cables must also have their fire performance specified, see Chapter 7. Note that some short lengths may be run outdoors for which there are various types of indoor/outdoor or 'universal' grade optical cables available from different manufacturers. These designs have adequate fire resistance for use within buildings and adequate weather resistance for use outdoors.

Other constructional details may be added where required such as tight-buffered or loose tube design.

Air blown fibre or cable is usually appropriate for the larger project where there is a significant saving to be realised by installing empty plastic tubes at the beginning of a project and then adding optical fibres as and when they are needed in the future. Air blown fibre should be considered against the following criteria:

- The 'larger' project.
- The cost and disruption to a business of having more cables installed at a later date is unacceptable.
- Concern about the correct selection of optical fibres and a decision is best deferred, e.g. OM3 versus OS1.

**Colour**   There is no universally accepted colour code for optical cables although there does exist ANSI/TIA/EIA-598-A-1995 *Optical Fiber Cable Color Coding*. Any colour can be made, of course, but this will usually mean a higher price, longer lead time and larger minimum order quantities. Outdoor and universal grade optical cables will nearly always be black because of the standard method of making cables that are stabilised against ultraviolet light.

## Optical cable and fibre standards to invoke

### Europe

- EN 188 000: *Generic specification for optical fibres.*
- EN 187 000: *Generic specification for optical fibre cables.*

### International

*Optical fibre*   Optical fibre is basically specified in the ISO 11801 2nd ed. series as OM1, OM2, OM3 and OS1. OS1 single mode fibre is fully described in ITU-T Rec. G.652. For those wishing additional optical and physical performance parameters then the following IEC standards apply:

- IEC 60793-1-1 (1999–02) Ed. 1.1: *Consolidated Edition Optical fibres – Part 1–1: Generic specification – General.*

- IEC 60793-2-10 Ed. 1.0: *Optical Fibres – Part 2–10: Product specifications – Sectional specification for category A1 multimode fibres.*
- IEC 60793-2-20 (2001–12): *Optical fibres – Part 2–20: Product specifications – Sectional specification for category A2 multimode fibres.*
- IEC 60793-2-30 Ed. 1.0: *Optical fibres – Part 2–30: Product specifications – Sectional specification for category A3 multimode fibres.*
- IEC 60793-2-40 Ed. 1.0: *Optical fibres – Part 2–40: Product specifications – Sectional specification for category A4 multimode fibres.*
- IEC 60793-2-50 Ed. 1.0: *Optical fibres – Part 2–50: Product specifications – Sectional specification for class B single-mode fibres.*

*Optical cable*

- IEC 60794-1-1 (2001–07): *Optical fibre cables – Part 1–1: Generic specification.*
- IEC 60794-2 (1998–08) Ed. 2.1: *Consolidated Edition Optical fibre cables – Part 2: Product specification (indoor cable).*
- IEC 60794-3 (2001–09): *Optical fibre cables – Part 3: Sectional specification – Outdoor cables.*
- IEC 60794-4-1 (1999–01): *Optical fibre cables – Part 4–1: Aerial optical cables for high-voltage power lines.*
- IEC 60794-2 (1998–08) Ed. 2.1: *Consolidated Edition Optical fibre cables – Part 2: Product specification (indoor cable).*
- IEC 60794-3 (2001–09): *Optical fibre cables – Part 3: Sectional specification – Outdoor cables.*
- IEC 60794-4-1 (1999–01): *Optical fibre cables – Part 4–1: Aerial optical cables for high-voltage power lines.*

**America**   ANSI/EIA/TIA-568-B.3 contains details of the optical fibre performances with Addendum 1 of part B.3 containing more information about 'laser' grade multimode fibre.

*Optical fibre*   More detailed information is contained in the following:

- ANSI/EIA/TIA-492AAAB: *Detail specification for 50/125 class 1a multimode, graded index, optical waveguide fibers.*
- ANSI/EIA/TIA-492AAAA-A-1998: *Detail specification for 62.5/125 Class 1a multimode, graded index, optical waveguide fibers.*
- ANSI/EIA/TIA-492CAAA-1998: *Detail specification for class IVa dispersion-unshifted single mode, optical fibers.*

*Optical cable*

- ANSI/ICEA S-83-596-1994: *Fiber optic premises distribution cable.*
- ANSI/ICEA-S-87-640-2000: *Fiber optic outside plant communications cable.*

## 12.5  Decide upon type of optical connector to be used

ISO 11801 2nd edition and the other standards recognise that there is now a large family of optical connectors available and the choice is larger than the SC connector originally specified in ISO 11801 1st edition. ISO 11801 2nd edition states that the optical connector at the TO shall be a duplex SC connector. The new standard goes on to say that connectors appearing at other locations may be any optical connector meeting an approved IEC standard and meeting the optical performance specified in ISO 11801 2nd edition.

It is not a major issue if users decide to go for other connectors at the TO other than the SC. The conformance to IEC standards and meeting the optical performance of ISO 11801 2nd edition is far more important. Users are recommended to go for small form factor connectors (SFF) where space is at a premium as they offer twice the packing density of traditional SC duplex connectors.

The connector decisions need to cover:

- *Appearance.* TO, CP, entrance facility, floor, building or campus distributor.
- *Multimode or single mode.*
- *Format.* simplex, duplex or SFF.
- *Type.* the common styles are:
  - ST
  - SC

    – FC-PC
    – LC
    – MT-RJ
    – FiberJack
    – VF-45(SG)
- *Termination style.* Users may select from the following or leave the choice at the discretion of the installer:
- Direct termination:
    – One of the adhesive/polish methods
    – Crimp and cleave
- Spliced preterminated tail cable.

Table 12.8 summarises the optical connector decisions.

Table 12.8 Optical connector project decision chart

| Appearance | Multimode or single mode | Connector style[1] | Termination method[2] | Fibre type[3] |
|---|---|---|---|---|
| **TO** Telecommunications outlet | | | | |
| **CP** Consolidation point | | | | |
| **FD** Floor distributor | | | | |
| **BD** Building distributor | | | | |
| **CD** Campus distributor | | | | |
| **EF** Entrance facility | | | | |

[1] Connector style: pick from, ST, SC-Duplex, FC-PC, MT-RJ, LC, Fiberjack, SG. This list is not exhaustive, other connector styles and variants are possible.
[2] Termination method: pick from, heat cure epoxy, cold cure, hot-melt, UV cured, fusion spliced pigtails, pre-stubbed/mechanical splice.
[3] Fibre type: pick from OM1, OM2, OM3, OS1. OM1 and OM2 should be suffixed with '50' or '62' to differentiate 50/125 and 62.5/125 options. OM3 is always 50/125. Fibre pigtails must be the same grade and size as the main cable.

## 12.5.1   Procurement guide

The following form of words is offered as a suggestion when specifying the cable system.

## *Optical fibre and cable*

An example optical cable specification might look like:

* *Item*: 1
* *Description*: Fibre optic cable
* *Quantity*: 7000 m

**Construction**   Non-metallic, multiple loose tube external grade optical cable suitable for immersion in water filled cable ducts. Two tubes must each contain eight individually coloured, primary coated multimode fibres and one tube must contain eight individually coloured, primary coated single mode fibres. The tubes will be sufficiently colour coded to allow unique identification of every single fibre in the cable.

**Specification**

* Diameter: 13-mm max.
* Finish/sheath: black polyethylene
* Operating temperature range: −20 to +70°C
* Tensile strength: 1500 N minimum
* Minimum bend radius: 175 mm
* Crush resistance: 500 N/cm
* Water penetration: IEC 60794-1

*Delivery and packing requirement*: Cables will be supplied as three times 2 km and one times 1 km on wooden drums with all exposed cable ends protected with waterproof heatshrink covers. The cable will be printed on the sheath, in white, "Optical fibre cable – external use only. 'Year of manufacture'".

### Optical parameters

- 16 off 50/125 µm and eight single mode primary coated optical fibres.
- Unless defined elsewhere, the optical fibre specification shall be according to IEC 60793 and more specifically the 50/125 shall be according to ISO 11801 2nd edition OM2 specification and the single mode shall be according to ISO 11801 2nd edition OS1 and ITU-T G652 specifications.

## *Patch panel termination and related items*

An example specification for the optical cable termination hardware could take the form of the following paragraphs.

### Optical patch panel

- 1U 19-inch rack-mounted optical patch panel, recessed by 100 mm.
- Finish to black painted steel.
- Front panel to be loaded with 16 multimode, beige, SC duplex adapters to IEC 60874-19-3 and eight, blue, SC duplex adapters to IEC 60874-19-2.
- The back of the panel shall have suitable cable glands to accept and support at least two optical cables in the diameter range 10–15 mm.
- The interior of the panel shall contain fibre management equipment to manage at least 24 one-metre pigtails including up to 24 fusion splice protector sleeves. The fibre minimum bend radius, as defined by the cable manufacturer for single mode fibre, will not be infringed at any time.

### Pigtail cable assembly

- 750–1000 mm of primary coated fibre to the same specification, and supplied by the same manufacturer, as the main cable.
- Termination at one end with SC multimode optical connectors to IEC 60874-19-1 and IEC 61754-4 or SC single mode connectors, to IEC 60874-14-5 and IEC 61754-4.

- All pigtails to be factory made and tested and individually packaged.

**Fibre optic splice protection sleeves**   Fibre optic heat-shrink splice protection sleeves shall be applied to all fusion splices. Sleeves shall be to any major telecommunication standard, for example, Bellcore, British Telecom and so on.

**Optical patch cord assembly**

- Each optical patch cord, simplex or duplex, shall be 1.5 m long and be constructed of tight buffered (900 μm) optical fibre to the same specification as the main cable and supplied by the same manufacturer.
- Patch cable shall be 2.8-mm (nominal) diameter with low flammability sheath (IEC 60332-1) and containing aramid yarn strength members.
- It will be terminated at both ends with SC multimode duplex optical connectors to IEC 60874-19-1 and IEC 61754-4 or SC single mode duplex connectors, to IEC 60874-14-5 and IEC 61754-4.
- All patch cords are to be factory made and tested and individually packaged.
- Patch cables must be of different colours to differentiate OM1, OM2, OM3 and OS1 optical fibres and the fibre type must be inkjet-printed onto the cable sheath.
- Duplex patchcords shall be cross-over, i.e. A to B and B to A.

# 12.6   Locate and specify the spaces

The 'spaces' are enclosed areas that house cables, equipment and terminating hardware. The spaces are an essential and integral part of the structured cabling system. The spaces include the telecommunications room, the equipment room and the building entrance facility. ISO 11801 2nd edition states that for every 1000 m$^2$ of office space there should be a telecommunications room.

Linking the spaces, and also the TOs, are the cable pathways.

These are defined routes that the cables take. The pathways may be simply marked out routes or more substantial cable management hardware. Chapter 8 deals with the spaces in more detail.

## 12.6.1    Essential design questions for the 'spaces'

### *Location*

Equipment and telecommunications rooms (TRs) must be located to take the following factors into account:

- *Distance from the majority of the work areas to be serviced.* The maximum horizontal cabling run allowed is 90 m. The TRs must be specially located to ensure this distance limitation is met.
- *Relative vertical position of the TRs.* Ideally the TRs should be placed one above the other in a multistorey building. This is to ensure that the backbone cabling is kept to a minimum and there are well-organised, accessible and spacious cable risers connecting them in the vertical plane.
- *Separation from sources of interference.* The rooms must be located well away from potential sources of electromagnetic interference such as lift shaft motors, electrical generating gear, television and cellular telephone transmitters, etc.
- *Accessibility.* The rooms must be located where it is possible to reach them with bulky and heavy equipment, e.g. a goods lift must be available with no intervening stairs.
- *The location must be environmentally secure.* For example, the rooms should not be below a water table if at all possible and must be impervious to all kinds of weather conditions.
- The building entrance facility must be located close to the actual point where the external cables enter and leave the building.
- The spaces must have direct access to the cabling backbone pathways with adequate space for all proposed cables while still maintaining all designated bend radii.
- Frames and cabinets must not be located in toilet facilities, kitchens, emergency escape ways, in ceiling or sub-floor spaces

or within closures containing fire hoses or other fire-fighting equipment.

- The spaces must not contain any other pipework, plumbing or electrical equipment that is not directly related to the operation of the telecommunications or cabling equipment within that space.

## *Other design features*

- *Security*. The rooms must be lockable and completely secure.
- *Accessibility to the rooms*. Doors large enough to wheel a two-metre equipment rack through must be available.
- *Power supply*. An adequate electrical supply must be supplied to the spaces and this must be presented with adequate mains power outlets for the equipment.
- *Uninterruptible power supplies, UPS*. It may be necessary to supply battery-backed UPS in the spaces.
- *Adequate floor space*. Sufficient floor space must be allowed for all of the equipment racks envisaged for the present and foreseeable future. This must include an allowance for the front access (at least 900 mm), other working areas, space for other equipment and UPS.
- *Adequate height*. The spaces must have sufficient height for the equipment racks to allow ventilation space and cable access above them. This means at least 2.6 m.
- *Adequate lighting*. There must be sufficient lighting installed that in all working areas it is possible to read black nine-point writing on a white background with ease.
- *Adequate floor strength*. The floor strength must be sufficient to cope with all of the envisaged equipment, including heavy UPS equipment with batteries.
- *Adequate ventilation and air conditioning*. If any active equipment is included in the spaces then they must be air conditioned with a temperature of 18 to 24°C easily maintained under all load and external environmental conditions. Battery storage systems must also be adequately ventilated.
- Building entrance facilities may require over-voltage protection equipment on copper cables entering the building.

- The earthing and equipotential bonding arrangements for all of the telecommunications equipment must be catered for.

## *Other control and monitoring features*

The following points should also be considered by/with the end user:

- Access alarms on the doors.
- Motion/infrared detectors within the spaces.
- Fire and smoke detection within the spaces.
- Fire suppression system within the spaces.
- Flood detection within the spaces.

## *The spaces checklist*

If all the following questions have been addressed then the cabling project requirements for the spaces will most likely have been met:

- Location.
- Floorspace.
- Height.
- Lighting.
- Air conditioning and ventilation.
- Floor loading.
- Interior layout of equipment.
- Access for heavy and bulky equipment.
- Local EMI/EMC environment.
- Power supply and UPS.
- Earthing and bonding requirements.
- Security.
- Fire, smoke and flood detection.

# 12.7    Determine the nature of the containment system

There is no standard that dictates the method of cable containment. The end user has a myriad of choices ranging from doing nothing to specifying a state-of-the-art cable trunking system. The choices will have to be made according to:

- How much money does the user want to spend?
- How much mechanical protection does the cable actually need in its environment?
- What value is given to the aesthetic appearance of the cable plant?
- What EMC/EMI protection is required?

The user can then choose from the following.

## 12.7.1   Cable routes marked out on a floor

Apart from nothing at all, this is the simplest form of cable containment. It can be appropriate when cable can be laid directly on a relatively smooth concrete floor which in turn is covered by a false floor.

## 12.7.2   Cable routes on the floor but with cable mat

Cable mat is spongy matting material that can be laid directly on rough concrete screed or wire basket tray to smooth out the mechanical load applied to points on the underside of the cables in contact with the floor or wire tray. Opinion is divided about whether this is really necessary or not. It would only be required on a concrete floor if the finish was extremely rough.

## 12.7.3   Cable conduit

Conduit is an enclosed tube that can be made of metal or plastic. It will provide maximum protection to the cable but will cost more to install and take longer to pull the cable into it.

## 12.7.4   Tray, trunking, wireway and raceway

Exact definitions of the above do not really exist, but their use implies a dedicated 'tray' that will hold the cables in place with easy access to those cables. Tray and trunking may be totally enclosed, but in which case the term 'duct' or 'ducting' is more often used, especially when the containment is permanently or semi-permanently enclosed. We will use 'cable tray' as the generic term for non-enclosed, rigid, cable support structures. The various forms and terminology are as follows:

- Ladder: so called because the construction looks like a ladder. It is used mostly in vertical risers for cable support but can be used horizontally as well. Cable ladder implies a larger construction that may be used for many cables and/or larger communications and power cables.
- Solid-bottom cable tray.
- Perforated or trough cable tray.
- Spine cable tray (centre rail construction).
- Wire tray (welded wire construction).
- Mesh cable tray (wire or plastic mesh).
- Wireway (fully enclosed with gasketed cover).
- Cable runway (no side panels; cable is tie wrapped to base tray).

## 12.7.5   Suspended cables

One method of containment is to suspend cables from hooks attached to the ceiling. These hooks are often referred to as 'J-hooks' as they look like a letter 'J'. This method is popular in the USA but not often seen in Europe. If the size of the cable bundle is not too large, there is nothing wrong with this method except that BS 6701 appears to forbid it with the statement 'no ceiling hangers' in its section 7.2.6.

## 12.7.6   Indoor cable duct

Underfloor duct systems are a cable management system embedded in the floor. The floor structure affects the type of underfloor duct system that can be accommodated in the floor and the total depth of concrete and method of pour will dictate the selection of the duct system. The various methods are:

- Monolithic pour.
- Slab-on-grade construction.

- Double-pour floor: the underfloor duct system is installed on the structural slab and the second pour buries the duct system.
- Prestressed concrete pour.

## 12.7.7  Room perimeter pathways

Room perimeter pathways are usually surface mounted trunking systems attached to the wall at desk level, sometimes called dado trunking, or at floor level called skirting trunking. These constructions are usually made from extruded white PVC and may be made up of several internal compartments. The internal compartments are for power and communications circuits. Some forms of trunking have metal separators between the compartments to improve EMC performance, and sometimes the plastic itself is metallised to achieve a similar, though not so effective, result. A common mistake is to pick very slim trunking for aesthetic reasons. However, when it comes to installing all the power and data cables there is not enough room. Similar considerations must be given to the use of power poles that have exactly the same function as the perimeter pathway. All power poles must be securely mounted so that they cannot be knocked over.

## 12.7.8  Firestopping

There may be local or national regulations that require the maintenance of a firestopping system where cable trays, conduit or trunking and so on pass through one fire zone to another. In such instances there will have to be some physical firestopping substance that stops flame or smoke or even the passage of air that may feed a fire further up the building.

## 12.7.9  Types of pathways – external

There are four types of external cable pathways:

- Underground cable duct.
- Direct buried.
- Aerial.
- Fixed to the outside of buildings.

### 12.7.10  Cable containment checklist

Locations that must be considered and specified for cable containment are:

- Under false floors
  - Marked out cable routes
  - Routes with cable mat
  - Conduit
  - Enclosed ducting
  - Open cable tray, various types.
- Above suspended ceilings
  - Conduit
  - 'J' hooks
  - open cable tray, various types.
- Vertical risers
  - Cable ladder.
- In-between buildings
  - Underground cable duct
  - Direct buried
  - Aerial
  - Fixed to the outside of buildings.
- Office space
  - Room perimeter raceways
      Skirting or dado height
  - Floor outlets
  - Power poles.

Chapter 8 gives more details of the cable containment options. Any firestopping requirements must also be considered.

## 12.8  Determine earthing and bonding requirements

Earthing, grounding and bonding covers the subject of ensuring that all exposed and other extraneous conductive surfaces are connected to earth and all cables screens are all effectively earthed. Chapter 9 gives more details on this subject.

All exposed conducting surfaces and other extraneous conductive parts must be connected to an earth point for safety purposes. This is to ensure that if, by accident, a live conductor touched these parts, the resulting circuit made to earth would cause such a large fault current that a fuse or a circuit breaker would blow somewhere. If the extraneous conductive part was not connected to earth, or left 'floating', then it would maintain the live voltage on it until somebody came along and touched it. Then that person's body would form the conductive path to earth.

In a structured cabling system the equipment racks, the active equipment, the metal patch panels and conduit, tray and trunking would all be considered as extraneous conductive parts and must be effectively earthed.

In a screened cabling system all of the screening elements of the cables, patch panels and connectors must also be earthed for the screening process to be effective. 'Floating' cable screens would be completely ineffective against interference and would also be considered to be a hazard as extraneous conductive parts.

The point of electrical bonding systems within buildings that use information technology equipment may be summarised as:

(a)  Safety from electrical hazards.
(b)  Reliable signal reference within the entire information technology installation.
(c)  Satisfactory electromagnetic performance of the entire information technology installation.

ISO 11801 2nd edition makes some comments about earthing and bonding and also states that it must be in accordance with IEC 60364 or applicable national codes.

## 12.8.1   Guide to specification

### Treatment of cable screens

Cabling screens, where they exist, must be properly bonded to earth for electrical safety and to optimise electromagnetic performance. All cabling components that form part of a screened channel should be screened and meet screening requirements. Cable screens shall be terminated to connector screens by low impedance terminations sufficient to maintain screen continuity necessary to meet cabling screening requirements. Suppliers' instructions on how to make low impedance terminations shall be requested and observed. Work area, equipment cords and the equipment attachment should also be screened and shall provide screen continuity when the horizontal cabling uses screened cable.

## *Earthing – general*

Earthing and bonding shall be in accordance with applicable electrical codes or IEC 60364-1. All screens of the cables shall be bonded at each distributor. Normally, the screens are bonded to the equipment racks, which are, in turn, bonded to building earth. The bond shall be designed to ensure that the path to earth shall be permanent, continuous and of low impedance. It is recommended that each equipment rack is individually bonded, in order to assure the continuity of the earth path.

The cable screens provide a continuous earth path to all parts of a cabling system that are interconnected by it. This bonding ensures that voltages that are induced into cabling (by any disturbances) are directed to building earth, and so do not cause interference to the transmitted signals. All earthing electrodes to different systems in the building shall be bonded together to reduce effects of differences in earth potential.

The building earthing system must not exceed the earth potential difference limits of 1-V rms (root mean square) between any two earths on the network. Corrective action must be taken with the earthing system to correct this situation if it arises.

## *Invoking standards*

In the United States users should invoke the following standard:

- ANSI/TIA/EIA-607: *Commercial Building Grounding and Bonding Requirements for Telecommunications.*

In the United Kingdom users should specify:

- BS 6701: *Code of practice for installation of apparatus intended for connection to certain telecommunications systems.*
- BS 7671: *Requirements for electrical installations, also known as the IEE Wiring Regulations 16th Edition and its corresponding Guidance Note No. 5 Protection Against Electric Shock.*

Users in the European Union should invoke:

- EN50310: *Application of equipotential bonding and earthing in buildings with information technology equipment.*
- EN 50174-2: *Information technology – cabling installation – Part 2: Installation and planning practices inside buildings.*

In any country the following may be used:

- IEC 60364-1: *Electrical installation of buildings – Part 1: Fundamental principles, assessment of general characteristics, definitions.*
- IEC 60364-4-41: *Electrical installation of buildings – Part 4–41: Protection for safety – Protection against electric shock.*
- IEC 60364-5-548: *Electrical installation of buildings – Part 5: Selection and erection of electrical equipment – Section 548: Earthing arrangements and equipotential bonding for information technology installations.*

## 12.9   The administration system

For a structured cabling system to maintain its value as an asset it is necessary to maintain accurate records of what the cabling consists of, what connections there are, what kind of numbering scheme is employed and to have an organised method of administering changes. All this information collected together is known as the administration scheme.

The method of recording this information may be varied. The choices are:

- Do nothing at all.
- Write down the connection details on a scrap of paper.
- Compile a neat hand-written workbook.
- Use a spreadsheet like MS Excel.
- Use a commercial database package like MS Access.
- Use a custom-made cable administration database.
- Use a real-time intelligent patch panel and database system.

Chapter 10 includes a discussion on when intelligent patch panel systems might be appropriate and cost effective. Intelligent patch panel systems invariably include the cable administration scheme as part of the package.

The user must decide what level of sophistication and detail he/she requires in the project and the likely cost. The cost of doing nothing must also be considered. The user must also decide who is going to input all of the original data into the database. Is it the installer or the users themselves?

It should be defined in the specification what level of cable administration scheme is required, what essential information needs to be recorded, how it will be recorded and handed over and who will enter the data in the first place.

## 12.9.1   The required data

The following minimum records regarding cabling infrastructure shall be provided:

- For cables: locations of end points, type, number, pairs.
- For outlets: identifier, type, location.
- For distributors: identifier, designation, type, location, connections.
- The floor plan, including the locations of the outlets, distributors and pathways.

### Optional records

When changes are made to the cabling infrastructure, including pathways and spaces, additional records may be necessary. Users should be wary of asking for too much information, however, as it is unlikely they will be able to cope with it all and, of course, there will be a cost involved.

**Cable records**   The following should be noted:

- Type of optical fibre or copper cable.
- Typical cable data (e.g. part number, sheath colour).
- Sheath and core identification.
- Manufacturer.
- Number of unterminated conductors and those with failures.
- Length.
- Data such as attenuation and crosstalk.
- Identification of pin connections at both ends and of splices.
- Performance classification (if applicable).
- Location of earthing.
- Treatment of screens.
- Transmission system under operation.
- Date code.
- Part number.
- Identifier.
- Linkage to identifiers for distributors, outlets, pathways and spaces.

**Telecommunications outlet records**   The following should be noted:

- Performance classification (if applicable).
- Single mode or multimode fibre (50/125 or 62.5/125).
- Screened or unscreened design.
- Manufacturer.
- Number and arrangement of terminated pins if not all pins are terminated.
- Part number.
- Identification of ports and cables connected.
- Linkage to identifiers for distributors, outlets, pathways and spaces.

**Distributor records**   The following should be noted:

- Number of available and used cables, fibres or pairs.
- Manufacturer.
- Number of conductors.
- Linkage to identifiers for cables, pathways and spaces.
- Part number.
- Front view of the patch cabinet.

**Pathway records**   The following should be noted:

- Type.
- Metal or non-metal design.
- Dimensions, mechanical data.
- Branching points.
- Manufacturer.
- Identification.
- Length.
- Location.
- Records of cables installed in that pathway.
- Location of earthing.

**Space records**   The following should be noted:

- Locations.
- Dimensions.
- Identification.
- Equipment located in the spaces.
- Space available.
- Type.

## 12.9.2   Presenting the data

If any computerised system is used to administer the cabling scheme, the presentation of the material to the end user will be in the format dictated by the particular software package in question. If using a paper-based scheme, some thought is needed to ensure that the

| **Location** Building 3, Canary Wharf | | | | | **Floor**    07 | | | |
|---|---|---|---|---|---|---|---|---|
| **Distributor number**    07-a | | | | **Rack number CC2\*** | 07-18 | | | |
| source | Cross-connect 1 | Cross-connect 2\* | CP | TO | Cable type | Fire rating | Link i/d | |
| 3-COM S-48-2 | 07A-14-11-09 back | 07A-18-10-23H | 07-18 | 07a-18-10-23T | C5U | EC-D | 07a-18-10-23 | |
| | | | | | | | | |
| Cisco A44-33 | -------------- | 07A-18-08-17H | ----- | 07a-18-08-17T | C6S | CMP | 07a-18-08-17 | |
| | | | | | | | | |
| Lucent FG98-8 | --------------- | 07A-18-13-48H MTRJ | ---- | 07a-18-13-48T SC-D | OM2-50 x 2 | HF-1 | 07a-18-13-48 | |
| | | | | | | | | |

| **Location** | | | | | **Floor** | | | |
|---|---|---|---|---|---|---|---|---|
| **Distributor number** | | | | **Rack number** | | | | |
| source | Cross-connect 1 | Cross-connect 2\* | CP | TO | Cable type | Fire rating | Link i/d | |
| | | | | | | | | |
| | | | | | | | | |
| | | | | | | | | |
| | | | | | | | | |
| | | | | | | | | |
| | | | | | | | | |

**Fig. 12.1** Example of a cable record sheet.

appropriate amount of detail is included. Figure 12.1 gives an example of a cabling administration document. Chapter 10 gives more explanation of what each field actually means.

## 12.9.3    Invoking standards

In the United States:

- ANSI/TIA/EIA-606: *Administration Standard for the Telecommunications Infrastructure of Commercial Buildings.*

In the European Union:

- EN 50174-1: *Information technology – Cabling installation – Part 1: Specification and quality assurance.*

In any country:

- ISO/IEC 14763-1: *Information technology – Implementation and operation of customer premises cabling – Part 1: Administration.*

# 12.10   Identify the civil engineering/rights of way issues for external cabling

If any cable goes between buildings then some thought needs to go into the cable routing.

## 12.10.1   Method

- Underground cable ducts.
- Direct burial in the ground.
- Aerial.
- Attached to buildings.

For all of the methods the following issues must be addressed:

- Who owns the rights of way?
- Are there any planning permission restrictions?
- For both underground and overground routes is there sufficient separation from other services, particularly high voltage cables?
- Do major obstacles have to be crossed, e.g. roads, railway lines, rivers, etc.?
- Do cable ducts exist? If so, who owns them? Is there space in them? Do you have permission to share them? Are they clear? Is there a draw rope already in place?
- If aerial cables are to be used, will they be self-supporting or attached to a catenary wire? Does the catenary wire exist? Do the telegraph poles exist? Is sufficient clearance going to be given to obstacles, crossings and high-voltage power lines?
- How will the external cables enter and leave the buildings? Is a building entrance facility required?
- Is overvoltage protection required? What specification will be

needed? Where will it go? Will it have any impact upon the cable performance?

- Where will the earth bonding be made for the metallic elements of a cable where it enters a building?

# 12.11  Installation instructions

Whereas most of the cabling specifications can be covered by various standards detailing specific items, products and their performances, there are numerous issues relating to the quality of the installation that the user would be well advised to specify at the tender stage.

## 12.11.1  General requirements

Some areas that might need to be included are:

- Cable separation. Power and data cables must be separated according to EN 50174-2.
- Use of cable ties. The vertical and horizontal positioning of the cable ties may need to be specified, as would their suitability. Over-tightened cable ties must not be used.
- Provision of dust caps on all optical connectors.
- Provision of seals on any blown fibre or cable ducts.
- Use of correct tools.
- Other installation details of EN 50174 should also be complied with.

## 12.11.2  Customer specific requirements

Users should define if there are any specific rules relating to operation on their site:

- No smoking policies.
- Health and safety obligations.
- Use of mobile phones and walkie-talkies in sensitive areas, e.g. airports, computer rooms, etc.
- Use of storage areas.

- Unloading areas and unloading times.
- Disposal of rubbish.
- Supply of electricity, water and sanitary requirements.
- Disabled access.
- Security requirements.
- Assumption of risk of delivered material.
- Insurance requirements, e.g. public liability, employers' liability and professional indemnity insurance.
- Confidentiality of customer's information.

# 12.12    Testing and project handover regime

At some point the cable installation needs to be tested to the customer's satisfaction and the project handed over to the end user.

## 12.12.1    Copper cabling

- What tests are going to be done? The appropriate test standard for ISO 11801 2nd edition is IEC 61935. For EN 50173 it is EN 50346. These standards list the tests that should be done. However they do not dictate what tests *must* be done. The user should invoke the standards and then reconfirm what tests are to be done.
- What is going to be tested? The user should define permanent link or channel tests. If a CP is being used, will this be one test 'through' the CP or two separate tests with the CP being at the ends? The permanent link test is usually the most appropriate for structured cabling and if a CP is used then a permanent link should still be tested as one unit incorporating the CP.
- What is the system being tested against? The test pass/fail parameters must be specified, e.g. ISO 11801 2nd edition Class E permanent link.
- What machine will do the testing? The user may have a preferred tester device; if not the level of capability must be specified, i.e.

at least a level IIe for Class D/Cat 5 testing and at least level III for Class E/Cat 6 testing.
- How many links will be tested? This can range from zero to 100%.
- How will marginal fails be treated? Marginal fails can be accepted or rejected. Owing to the accuracy of the test machines it is acceptable, and the most economic all round, to agree to accept all marginal failures.
- How will the test data be handed over to the customer? The test data handed over might be nothing at all, paper copies, a paper summary or electronic files. The format of electronic files can vary. A user might want them in .txt or .doc format so that anybody can open or read them. However, they cannot be mass processed in this format. If there are 50 000 outlets in a project this amounts to 50 000 test files or A4 sheets of paper. It would be impossible to study them all. In such a case the information should be stored in the file format accompanying the tester itself, e.g. if using a Fluke DSP4x00, the information should be recorded in .fcm files. The cable manufacturer should also be consulted if there are warranty implications concerning the method of testing and recording the information.

## Example test specification for a Category 6 cabling system

1   The copper cable system will be tested in accordance with IEC 61935.
2   The tests will include, as a minimum, the following:
    - Wire map
    - Length
    - Attenuation
    - NEXT
    - PS-NEXT
    - ELFEXT
    - PS-ELFEXT
    - DC resistance
    - Return loss

- Delay
- Skew.

3   The cabling section tested will be the permanent link as defined in ISO 11801 2nd edition.

4   The test parameters will be set to ISO 11801 2nd edition Class E permanent link.

5   The link will be tested from both ends with a suitable tester meeting the Level III requirements.

6   100% of the cable plant will be tested.

7   Results indicated as a 'marginal fail' by the tester will be accepted. Links marked as an outright fail by the tester will be repaired and retested before they will be accepted by the customer's representative.

8   The customer's representative retains the right to witness any or all of the tests. The test schedule will be communicated to the customer's representative at least 48 hours in advance of the tests.

9   A paper-based summary sheet of the tests will be handed over as part of the project documentation. An electronic copy of the tests, on a CD or floppy disc, will be handed over as part of the project documentation. Two copies are required. One with a .txt file extension and one with the file format of the testers own cable management programme, e.g. .fcm or .ltrm.

10   The correct Category 6 interface will be used with the tester as specified by the cable manufacturer. The interface will be replaced at an interval specified by the tester manufacturer. Only generic test software may be used, i.e. not specific to any cable manufacturer.

11   All test equipment will be calibrated according to the test equipment manufacturer's recommendations and all operators will have undergone appropriate training.

## 12.12.2   Optical cabling

Optical cable can be tested with an optical power meter/light source combination or with an OTDR. Unless the cable link is more than 2000 m in length, the author strongly recommends that only the power meter method is used for acceptance testing. As most structured cabling systems are less than 2000-m long, this would seem to preclude the use of an OTDR completely, and this is correct. An OTDR should be used for fault finding and characterising very long lengths of optical fibre. An OTDR is inappropriate for short length acceptance testing because:

- Unless it is used correctly, i.e. with long launch and tail leads, short pulse lengths etc, it is very difficult to 'see' the fibre in question or the connectors on the ends.
- The OTDR will not automatically give the attenuation of the link under test, it can only be inferred from the trace.
- Experience shows that less than half of data cable installers are really proficient at using an OTDR.
- The resulting OTDR trace requires expert interpretation; who is going to do that? One project that the author worked on yielded 64 000 OTDR traces!
- All the above add up to unnecessary cost and unreliable handover documentation.

So unless the project has fibre lengths in excess of 2000 m or unless a fault is being sought, a power meter/light source test should be used.

## Example test specification for multimode optical cabling

1   The optical cable plant will be tested with an optical power meter and stabilised light source according to EN 50346 or IEC 61280.
2   The link will be tested in both directions at 850 nm and 1300 nm.
3   Method 1 of EN 50346 is the recommended method. If this cannot be achieved owing to mismatch of connector types and/or the use of an automated optical tester, then Method 2 of EN 50346 may be used. The test method must be explained to the satisfaction of the customer's representative and if necessary the cable system manufacturer.
4   All test equipment will be calibrated according to the test equipment manufacturer's recommendations and all operators will have undergone appropriate training.

5   Index matching gels will not be used to improve the test results.

6   A fibre link will be accepted as having passed if the attenuation at 850 and 1300 nm, and when measured in both directions, is equal to or less than the link attenuation (for 300, 500 and 2000-m links as appropriate) published in ISO 11801 2nd edition. Links not meeting this requirement will not be accepted and must be repaired and retested.

7   The customer's representative retains the right to witness any or all of the tests. The test schedule will be communicated to the customer's representative at least 48 hours in advance of the tests.

8   A paper-based summary sheet of the tests will be handed over as part of the project documentation. The full test results will be recorded on a handwritten form as typified by Fig. 11.13 for power meter – light source results, or in an electronic file format when automated test equipment is used.

# 12.13   Prequalification and bidding

The final part of writing the invitation to tender (ITT) or request for proposal (RFP) is to consider a little more deeply the companies who will be tendering for the business. The kinds of companies that a user will be dealing with will include:

- Cable system manufacturers.
- The larger distributors.
- Installers.
- Consultants.

The role of the consultant may be to help in defining and designing the cable system, writing the proposal, organising the prequalification and judging the tenders and making selection recommendations to the end user client.

Prequalification is becoming a more popular step in the purchasing cycle, especially for larger projects. It helps prevent the end user from being swamped by large numbers of low quality bids. Prequalification may include the cable manufacturer and the installer as well. The process usually involves asking the potential supplier a number of questions about their product range, the technical performance they can offer, what standards they meet, what quality systems they run, what kind of warranty they can offer and what track record can they show.

The result of this exercise will allow the user typically to select about six manufacturers and about four installers. Each of the four installers is then allowed to bid any of the six prequalified manufacturers' systems.

In this chapter we have seen that the system design and tender writing process involves making some fundamental decisions about the kind of technology to be used, the topography of the cabling system, the test regimes required and several other technical issues. The technical part of the tender can describe exactly what is required, where it is to go and how it must be tested. The user may, however, decide to take a more distanced approach and allow the installer to suggest products and designs. Either way there are a few more questions that should be asked of suppliers, whether in the prequalification or final selection phase. These can be summarised below:

For the manufacturer:

- Summarise the product range.
- What international standards can you meet?
- What is the technical performance of your main products?
- Can you demonstrate third party compliance certificates for your products and overall system?
- What quality systems do you have?
- What environmental system does your company operate?
- What exactly is covered by your system warranty, e.g. are EMC/EMI requirements included?
- What does an installer have to do to get the warranty from you?
- What happens in the event of a claim?
- What warranty claims have you had?
- What happens if the installer goes out of business?
- What does an installer have to do to become an authorised reseller of your products?

- What reference sites do you have of a similar size, in a similar industry, in the same area?
- How long have you been supplying these products?
- Are you the original manufacturer or a reseller?
- What other services can you offer, e.g. training, on-site technical support, monitoring of the installer etc?

For the installer:

- How long have you been in business?
- What turnover has your company achieved over the last three years?
- How many employees do you have?
- What reliance do you have on subcontract labour?
- How do you supervise and train the subcontract labour?
- What reference sites do you have of a similar size, in a similar industry, in the same area?
- How long have you been supplying these products?
- What is your relationship with the principal manufacturer?
- How many manufacturers do you represent?
- What test equipment do you own?
- Describe the training regime you have in place for installers, testers and supervisors?
- Who decides upon and signs off the final design and what are their experience and qualifications?
- What quality system does your company operate?

# 13

# Example specification and design

## 13.1 Introduction

In this final chapter we shall consider the design of a structured cabling system for a small campus project consisting of three buildings, each with four floors.

Let us call it the University of Cheshire, Shiloh Park campus. It is a modern development, custom built for the university but with an eye on the future in case a private industry might spring up in the midst of academia. The developers have partly funded some of the project and maintain an option on some of the office space.

There are three main buildings: the largest is the principal office block called the Hexagon Centre, for obvious reasons. The two other buildings are conventional laboratories and lecture theatres with a few seminar rooms and offices. They are called the J. L. Jackson Building and the Humphrey Laboratory. The latter two buildings will need a fairly low density of cabling and outlets but the main office block, the Hexagon Centre, will be very densely cabled.

We approach the project in the later stages of construction at the final fit out. As usual in large construction projects the data cabling has been left to the last minute and now the main contractor needs a fast response to some very basic design guidelines. We will place ourselves in the position of the installer (contractor) with the brief to come up with a simple yet comprehensive structured cabling system design with preliminary costs.

## 13.2   The prime contractor's brief

### 13.2.1   Work package 871 – structured cabling – building and interbuilding

Grovis Construction plc, acting on behalf of the Clerk of Works of the University of Cheshire, invites preliminary design and budgetary costing for the structured cabling works at the University of Cheshire, Shiloh Park Campus project.

There are three buildings on the campus: an office block called the Hexagon Centre and two mixed laboratory/teaching buildings called the J. L. Jackson Building and the Humphrey Laboratory. The general layout is shown in Fig. 13.1. This is not a scale drawing but is sufficiently accurate for the preliminary design budgeting exercise. Contractors must take off approximate scaled distances from these drawings. Contractors invited to bid for the full work package will be allowed to survey the site fully in the near future.

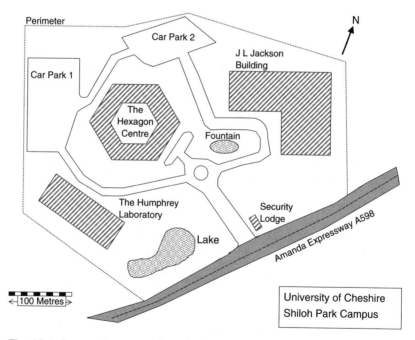

**Fig. 13.1** Overall plan of Shiloh Park Campus.

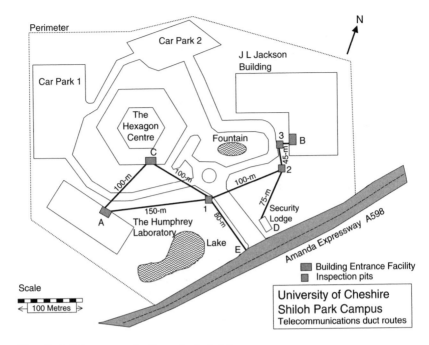

**Fig. 13.2** Telecommunications duct routes.

Underground cable ducts have already been laid and their specification is given in Fig. 13.4. The routes of the ducts and the disposition of maintenance and hand holes are given in Fig. 13.2. One duct is reserved for British Telecom who will lay their own cable from the roadway (The A598 Amanda Expressway) into the main computer room in the Hexagon Centre.

The architects have already designed in a main computer room (as shown in Fig. 13.3) on the ground floor of the Hexagon Centre and on each of the four floors, two telecommunications rooms have been designated for information technology equipment. Vertical risers, consisting of eight 100-mm vertical ducts, link these rooms.

The building is four stories high, made of a steel frame with concrete cladding. The floors are steel pans with poured concrete finish. The client has designed the Hexagon Centre to be extremely flexible in the future use of floor space. To this end each floor has been divided up into a number of standard office module designs, forming large open-plan spaces between the main load-bearing walls. The

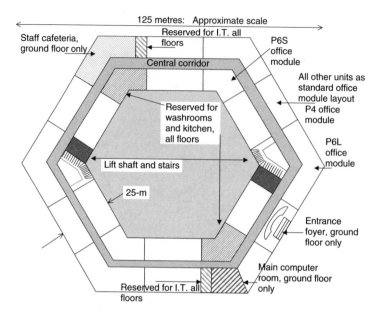

**Fig. 13.3** General floor plan of the Hexagon Centre building.

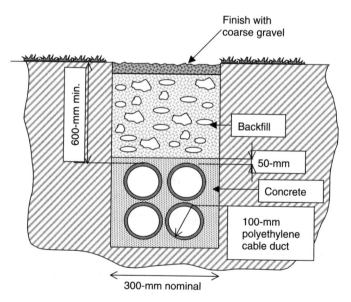

**Fig. 13.4** Typical telecommunications duct specification.

office spaces will be let as open plan but with the easy ability to add in simple stud partition walling if desired.

The office units are known as a polygon-4 unit (P4), basically a rectangle 21 by 9 m; a polygon-6 large unit (P6L), a large irregular six-sided polygon on the outside of the building and a polygon-6 small unit (P6S), a smaller irregular six-sided polygon on the inside of the building.

Certain areas are reserved for stairwells, lift shafts, toilet and kitchen areas. On the ground floor one P6L section is reserved for the staff cafeteria and another area is designated for the main computer room. None of these areas need structured cabling. Also on the ground floor one P4 unit is taken up with the main entrance foyer. Twelve outlets need to be terminated at the main desk for telephones and computers of the reception staff.

A central corridor, 3 m wide, circumnavigates the entire building on each floor. Floors 1, 2 and 3 (the convention used is 'ground, first, second and third' floors) are identical in layout.

All office area flooring is raised giving a 30-cm clearance. The corridor floor is solid concrete raised to a height to give an even transition height into each of the office areas. The whole building has suspended ceiling giving a 30-cm clearance beneath the concrete floor deck above.

## 13.3 The client's brief

The end user of the network ('the client') is the University of Cheshire. They have laid down the following specification:

1 The horizontal cabling system shall be based on two copper data/telephone outlets per user position. Every predicted or actual work station area shall be within 3 m of a pair of outlets.
2 The cable shall meet the latest international standards. The grade and type of cable shall be suitable so that it has a predicted life of at least six years before any major upgrade is required.
3 A manufacturer's warranty is required of at least 15 years' duration.

4    No special EMC/EMI problems are envisaged so it is expected that an unscreened cable solution will suffice.
5    A low fire hazard material is required for the cable sheath.
6    An optical fibre backbone is required between floors, main IT rooms and between buildings.
7    A 100-pair telephone cable is to be installed parallel to the fibre optic backbone cables.
8    The system must be designed to allow at least gigabit Ethernet to every desk position, analogue or digital telephony to every desk position and at least ten gigabit Ethernet transmission in the backbones.
9    A copper telephone cable must be laid to the security lodge.

## 13.4    The contractor's response

We must take the step-by-step design approach discussed in Chapter 12.

- First pick the standards philosophy.
- Decide upon the overall topology and density of outlets.
- Decide upon category of cable and class of optical fibre.
- Decide upon type of optical connector.
- Decide upon unscreened or screened copper cable.
- Decide upon fire rating of internal cables.
- Locate and specify the spaces.
- Determine the nature of the containment system.
- Determine earthing and bonding requirements.
- Decide upon the administration system.
- Identify the civil engineering/rights of way issues for external cabling.
- Provide specific installation instructions.
- Decide upon the testing and handover regime.

### 13.4.1    Standards philosophy

We shall base the design on ISO 11801 2nd edition but we can invoke other CENELEC and IEC standards as necessary.

## 13.4.2   Overall topology

The project looks fairly conventional with a three-level hierarchy: the horizontal cabling, a building backbone between the telecommunications rooms and a campus backbone between the buildings. No request had been made for cross-connect or CPs so the simplest and cheapest solution will be the two-connector model for the horizontal cabling. The number of outlets has not been specified but it is an open-plan floor-grid system. We can calculate the number of outlets from a simple scale diagram. No mention is made of FTTD or COA so maybe these can be mentioned as optional alternatives.

## 13.4.3   Decide upon category of cable and class of optical fibre

The RFI (request for information) asks for cable to the latest standards, suitable for at least gigabit Ethernet and lasting at least six years. Cat5e/Class D could achieve this but the request for a six-year life before upgrade, plus the size and prestige of the building makes this look more like a Category 6/Class E project. It would be wise to mention in the response though that Class D could be offered at a slightly lower price but with a much lower specification.

The class of optical fibre would appear to come down to OM3 and/or single mode fibre to support 10-gigabit Ethernet across the campus. We can see from Fig. 13.2 that the longest link is from the Humphrey Lab to the J. L. Jackson Building, A-B, 315m. This is just outside the range of the 10-gigabit Ethernet, which is guaranteed to transmit up to 300m on OM3 fibre.

There is not enough information in the RFI to state specifically how many fibres are necessary so one might as well pick the average number of eight. We would select eight OM3 and eight single mode (OS1) fibres to go between each of the buildings and eight OM3 fibres within the buildings where required.

An ideal solution might be a special outdoor duct cable containing, say, eight OM3 and eight OS1 fibres, but this is unlikely to be a stock item at any distributor and the total quantity required, around 680m would make it uneconomic to be made as a 'special'.

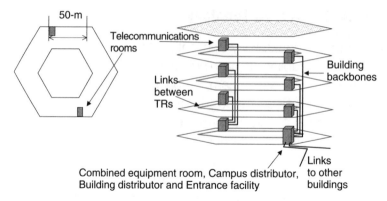

**Fig. 13.5** Building backbone for the Hexagon Centre.

Another solution is 680 m of a duct grade 8-fibre OM3 cable and 680 m of 8-fibre OS1 cable. The cables would be run from C to A, A to B and C to B. This means that in the duct route labelled 1,2,3, the cables are doubling up. It would be clever to put some sort of flexibility or distribution point in inspection pit number 1. But for the short lengths involved it is probably uneconomic to do this and cheaper to run the two cables across the site in continuous lengths.

A third option is to put in tubes for air blown fibre. A seven-way multiduct is potentially an 84-fibre cable and an external grade multiduct could be pulled in the same duct routes across the campus. Eight OM3 fibres could be blown into one tube and eight OS1 into another, leaving five empty ducts for expansion.

Within the Hexagon Centre we would run four-way multiduct around the corridors to link the two telecommunications rooms (TRs) on each floor to give circuit resilience. We would also link each TR on each floor back down to the main equipment room with another four-way multiduct. We could then blow in eight OM3 fibres around one of the multiduct tubes. Figure 13.5 shows the building backbone structure for the Hexagon Centre.

## 13.4.4   Decide upon type of optical connector

Remember that optical connectors can appear at several locations within the cabling system and there is no rule that says we must use the same connector in all places.

There is no stated requirement for fibre-to-the-desk so we must consider only the floor, building and campus distributor locations. Space does not appear to be a great issue so there is no particular reason to specify small form factor connectors. We may as well pick the 'standard' connector, that is the SC connector. We will have to select single mode and multimode versions for the various fibres that we have in mind. Remember that there are only multimode and single mode connectors; we do not have to specify a different connector for OM1, OM2 or OM3.

We shall specify that all multimode fibres will be directly terminated on-site, for lowest cost, but all single mode fibres will be terminated with factory-made pigtails that will be fusion spliced onto the main cable on-site.

## 13.4.5   Decide upon unscreened or screened copper cable

The customer appears already to have settled for unscreened cable and there is no other information available which would appear to need to argue any differently.

## 13.4.6   Decide upon fire rating of internal cables

The client has made some comments about low fire hazard cables but it does not appear to be a major issue as no standards have been invoked. It would probably be best to offer the lowest price of the zero halogen, fire retardant cables meeting the IEC 60332-1 specification. It would be best to advise the client that cheaper forms of PVC cabling are available as are more expensive, but higher performance products, meeting IEC 60332-3-24-c or even plenum grade UL 910.

## 13.4.7   Locate and specify the spaces

If we consider the plans for the Hexagon Centre we can see that the main spaces are already specified. There appears to be a combined equipment room, TR and building entrance facility (acting as floor, building and campus distributor) on the ground floor. In the opposing diagonal there is another TR and the pattern of two TRs on each floor is repeated on each of the other three floors.

The client has stated that each TR is placed one above the other and linked by vertical risers consisting of eight 100-mm ducts. The two TRs on each floor must also be linked to each other.

All of the building, except for the central corridor, appears to have false floors and ceilings. The central corridor has a solid concrete floor but a suspended ceiling.

Across the site we can see that cable inspection pits are available but we do not know their size.

The Humphrey Laboratory has only one specified space and we will have to work around that and come up with suggestions of our own. The disposition of the J. L. Jackson building is left as an exercise for the reader.

We can work out the approximate size of the spaces from the diagrams and we will return to the suitability of the available space after the first draft of the overall cabling design to check if the spaces are large enough.

As no further details are given we must state our assumptions and requirements concerning the spaces when we submit our draft proposal.

## 13.4.8   Determine the nature of the containment system

No instructions are given concerning the containment system so we should propose something suitable and economic.

We propose wire basket in the suspended ceiling spaces and cable laid directly on the floor in the false floor areas. As this is a new building the concrete floor should be in a good enough condition. The installer should, however, point out that a simple survey of the floor quality would be needed to confirm this. Options can be given to the client of laying cable matting or even wire basket across the floors.

New cable ducts are in place across the campus but once again the installer must state their expectations about availability, existence of draw ropes and so on.

### 13.4.9   Determine earthing and bonding requirements

Other information received from the prime contractor indicates that the electrical supply and protective earth system will form the basis of another contract.

As this project will use unscreened cable and there is no other known requirement for functional earths then earthing requirements will be fairly simple.

However, it is still the responsibility of the installer to:

- Earth all equipment racks, including doors, and the equipment contained within them.
- Earth all metal containment systems.
- Earth any armour or other extraneous metallic element of any cable entering a building.
- Supply overvoltage protection devices to metallic cables entering a building.

### 13.4.10   Decide upon the administration system

A logical numbering scheme will be adopted and applied to all cables ensuring that every cable has a unique identity. A proprietary labelling and marking method will be used to generate labels to go on all TOs, patch panels and the cables themselves. The client will be given the option of being given this information in an Excel spreadsheet or being loaded into a proprietary cable management database programme.

### 13.4.11   Identify the civil engineering/rights of way issues for external cabling

There should be no rights of way issues as all of the cabling is contained within the campus and in ducts belonging to the university. One duct will be reserved for BT cables running from the main road into the building entrance facility within the Hexagon Centre.

As mentioned in the section on cable containment, the accessibility and suitability of the ducts must be confirmed by the prime contractor. Surveying and possible rodding and roping of the ducts if they are blocked or without a draw rope will be an expensive business.

### 13.4.12  Specific installation instructions

There are no specific installation instructions that we are aware of at this stage.

### 13.4.13  Test and handover regime

No particular test regime is specified so we shall propose a 100% test plan as defined in Chapter 11.

## 13.5  The first draft design

### 13.5.1  Location of the spaces and the horizontal cabling

The first step is to do a basic sizing exercise to ensure that the location of the TRs will still allow the 90-m horizontal cabling limit to be applied.

Figure 13.6 shows a scale diagram of one of the floors. Assuming the cables will leave the TRs via the false ceiling above the main corridor, we can see that the cable will have a maximum run of about 77 m to the far end of the furthest office. Even this can only be achieved if the cable is laid in a straight line across the ceiling of the last office unit. A length of 77 m is less than the statutory 90 m but it is unwise to design any horizontal link with more than an apparent 75 m point-to-point run. The remaining 15 m will always be consumed by cable drops from the ceiling to the outlet, going around unknown obstructions and the inherent tolerance that needs to be built in to any simple measuring exercise such as this.

From Fig. 13.6 we can also conclude that we will have to service the offices by forming two zones. Zone G2 will be served from telecommunications room TR G2 and Zone G1 will be fed from

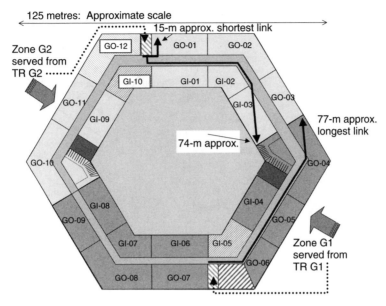

**Fig. 13.6** Forming two zones from TR G2 and TR G1.

telecommunications room TR G1. The designation 'G' used here represents the ground floor.

Another figure to arrive at is the average cable run. If we take the average of the shortest and longest cable runs we will have a remarkably accurate estimate of the total cable requirement without having to work out laboriously thousands of scaled routes drawn on the diagram. The longest link is 77 m and the shortest is 15 m. Thus the average cable length is 46 m. This figure is remarkably constant throughout the industry.

We must now look at the TOs in each of the three basic types of open plan office unit. The three units are expanded in Fig. 13.7. The client's specification called for a pair of outlets to be within 3 m of any prospective user or workstation.

We suggest a grid based on two rows of floor outlets plus twin outlets set in dado trunking around the wall. The floor boxes will be standard floor outlets containing four Category 6 outlets and some power outlets. The dado rail will be a standard white PVC three-compartment trunking with double Category 6 outlets set into it at

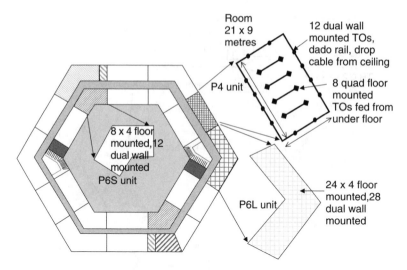

**Fig. 13.7** Telecommunications outlets in the three basic unit floor plans.

intervals along with power outlets. We must remember the constraints imposed upon us concerning proximity of power and data cables by EN 50174.

The floor boxes will be fed by cabling laid along the floor laid directly on the ground. The wall outlets will be fed from wire basket tray secured within the ceiling space. All the cables will have to travel from the TRs along the ceiling spaces as the main corridor has a concrete floor, so each room will need a large section of riser trunking to get all the floor cables from the ceiling into the floor space.

If we look at the rectangular unit called P4 we can see eight quad floor outlets and twelve double wall outlets will give the required density and distance from prospective users. This gives a total of 56 outlets in $189\,m^2$ of floor space, or one pair of outlets per $6.75\,m^2$ of floor space. This is well within the two outlets per $10\,m^2$ of floor space that is the accepted minimum density for ordinary office space. We can also estimate from the diagram that no user is more than 3 m from an outlet.

A similar design exercise on the P6L unit reveals 152 outlets and then 56 outlets for each P6S unit. The reader may wish to fill in the design sketches themselves to confirm this. We are now in a posi-

Table 13.1 Total number of office units in the Hexagon Centre

| | Office unit | | | |
|---|---|---|---|---|
| | P4 | P6S | P6L | Foyer |
| Ground floor | 11 | 2 | 4 | 1 |
| 1st floor | 12 | 2 | 6 | |
| 2nd floor | 12 | 2 | 6 | |
| 3rd floor | 12 | 2 | 6 | |
| Total | 47 | 8 | 22 | 1 |

Table 13.2 Total number of outlets required

| | Office unit | | | |
|---|---|---|---|---|
| | P4 | P6S | P6L | Foyer |
| Ground floor | 11 × 56 = 616 | 2 × 56 = 112 | 4 × 152 = 608 | 1 × 12 = 12 |
| 1st floor | 12 × 56 = 672 | 2 × 56 = 112 | 6 × 152 = 912 | |
| 2nd floor | 12 × 56 = 672 | 2 × 56 = 112 | 6 × 152 = 912 | |
| 3rd floor | 12 × 56 = 672 | 2 × 56 = 112 | 6 × 152 = 912 | |
| Total | 2632 | 448 | 3344 | 12 |
| Grand total | 6436 | | | |

tion to start estimating the quantities of main components required to cable up the Hexagon Centre.

Table 13.1 lists all of the various kinds of office unit and how many of them there are, not forgetting the foyer area with its requirements for 12 outlets. Table 13.2 multiplies all of the office units by the number of outlets allocated to each type. This gives a grand total of 6436 outlets.

The total of 6436 outlets with an average cable length of 46 m gives a total cable requirement of 46 × 6436 = 296 056 m. We need to add to this about 5% for measurement tolerance and scrap to arrive at 310 858 m. If this comes in 305-m boxes this amounts to 1020 boxes (29 pallets!) or 622 500-m cable drums. Obviously we must remember to specify some secure and dry storage areas needed in this project!

Again 6436 outlets will require 6436 divided by 24, patch panels, that is 269 24-port panels. It is actually more than this because, as seen in Table 13.3, the rounding-up effect in each TO means we actually need a total of 273. For every two panels we should allow a 1-U cable management bar. So 273 panels will require about 360 U of rack space. Using 40-U racks we can see that we will need nine 40-U racks. For as many racks dedicated to cabling we can assume we will need just as many for the active equipment.

This is a good point to pause for another sizing check. The preceding calculation shows that about 848 copper cables will need to be terminated in each TO. This requires 36 patch panels and 12 cable management bars, or 48-U of space and needs two racks plus two more for the active equipment. The footprint of an average rack is about $0.94 m^2$, made up from the depth of 660-mm plus 900-mm front access times the width, 600 mm. So a minimum of $4 m^2$ will be needed for racking. We can double this to allow for PABX equipment and terminations, UPS power and other related IT equipment plus some basic working office space. This comes to a minimum requirement of $8 m^2$. By checking the scale drawings it appears that about $40 m^2$ have been set aside for each TO. This is more than enough and unusually generous. This allowance is unusual. More often than not the opposite is true with hopelessly inadequate space devoted to IT requirements.

Another check to do is to look at the volume of cabling building up in the ceiling containment and thus estimate the size of cable tray required. The worst case will be in the corridors approaching the TOs. If there are 848 cables per TR we can assume that 424 will approach from each direction in the corridor ceiling space. The 424 Category 6 cables will occupy $424 \times 36 mm^2$, that is $15264 mm^2$. With a 50% fill allowed in trunking this comes to about $30000 mm^2$. So the cable tray needs to have dimensions along the lines of $500 \times 60 mm$. However, if the cables are in bundles of 24 they will have heights of about 36 mm per bundle, sketching this out shows that two layers of such cable bundles will have a maximum possible height of 72 mm. As we do not want cables to exceed the height of the cable-tray walls, 60 mm would not be high enough. The trays would have to be the nearest standard size larger than $500 \times 72 mm$.

We have already decided that the optical fibre building backbone is going to be four-way air blown fibre multiduct. Figure 13.5 demonstrates the building backbone routes and the inter-TR links.

Each TR on the ground floor will have five cables (bloducts) entering them. Five four-way bloducts could amount to $5 \times 4 \times 12 = 240$ fibres. 12-U of rack space should be reserved to accommodate this in future expansion. However, with our intention of blowing eight fibres per route on the first day (day one) we only expect to terminate 40 fibres.

## 13.5.2    The building backbone

TR G2 is also the campus distributor so it will receive two seven-way multiducts from the campus backbone. The potential volume of fibre to be terminated is $7 \times 2 \times 12 = 168$. Assuming 24 fibres terminated per 1-U of rack space, plus 33% extra for cable management, means 10-U should be reserved. The day-one design, however, calls for eight OM3 and eight OS1 fibres in each multiduct, giving 16 multimode and 16 single mode terminations.

We can estimate the route lengths from Fig. 13.5. A complete circuit of the main corridor is about 300 m. The four floors will thus require 1200 m of four-way microduct.

For the riser links we should allow 5 m per floor. This adds up to 60 m for the vertical links for both sets of TOs. Total volume is therefore 1260 m plus 5%, giving 1323 m. As manufacturers tend to sell this in 500-m lengths we will need to cost this as 1500 m.

We have yet to mention the 100-pair copper telephone cable. This is presumably of Category 3 standard but in most countries it will be sold according to the local telecommunications supplier's standard, for example British Telecom CW1308 in Britain.

Telephone cables are categorised as indoor or outdoor (usually gel-filled) grades, number of pairs and size of the copper wires. There might also be an earth wire included in the cable to give a functional earth connection for some types of PABX. The sizes of the copper wires are given by their diameters in millimetres and are 0.4, 0.5, 0.6 and 0.9 mm, respectively. The larger the copper wire the lower the attenuation and the further the signal will go, but the cable will cost more and be of larger physical size. A diameter of 0.4 mm is probably good enough for the distances to be covered in this example project, but if the object is to use the telephone cable for some future digital communications service, such as voice over IP, then it would be best to get as close as possible to the Category 3 standard and this means a 0.5-mm conductor.

# 13.6    The bill of quantity

We are now in a position to start adding up all the components that we need to complete the project for the campus cabling and the Hexagon Centre. Table 13.3 allows us to list all the items required in the TRs and the campus distributor/equipment room. The list is not, of course, exhaustive but does contain all the principal items needed for a structured cabling system.

## 13.6.1    Item 1: category 6 cable

- *Quantity*: 310 km, to be supplied as 1020, 305-m boxes or 622, 500-m reels.

Table 13.3 Telecommunications room equipment list

| TR | C6 cables | 24 port panels | Fibre terminations | | | | | Cable mngmnt bars | Total 'U' required | 40-U racks required |
|---|---|---|---|---|---|---|---|---|---|---|
| | | | Multi-mode | Single mode | Fibre potential | Actual fibre panels needed | Fibre panels possible | | | |
| TR G1 | 680 | 29 | 56 | 16 | 408 | 3 | 17 | 16 | 62 | 2 |
| TR G2 | 668 | 28 | 40 | | 240 | 2 | 10 | 13 | 51 | 2 |
| TR 11 | 848 | 36 | 24 | | 144 | 1 | 6 | 14 | 56 | 2 |
| TR 12 | 848 | 36 | 24 | | 144 | 1 | 6 | 14 | 56 | 2 |
| TR 21 | 848 | 36 | 24 | | 144 | 1 | 6 | 14 | 56 | 2 |
| TR 22 | 848 | 36 | 24 | | 144 | 1 | 6 | 14 | 56 | 2 |
| TR 31 | 848 | 36 | 24 | | 144 | 1 | 6 | 14 | 56 | 2 |
| TR 32 | 848 | 36 | 24 | | 144 | 1 | 6 | 14 | 56 | 2 |
| Totals | 6436 | 273 | 240 | 16 | 1512 | 11 | 63 | 113 | 449 | 16 |

- *Specification*: four-pair, 100-Ω characteristic impedance, un-screened, meeting the Category 6 electrical performance of ISO 11801 2nd edition and IEC 61156 or EN 50288. Colour, grey. Sheath fire performance to meet IEC 60332-1, IEC 60754-1 and IEC 61034.

## 13.6.2   Item 2: telecommunication outlets

- *Quantity*: 6436
- *Specification*: eight-position 'RJ-45' socket to IEC 60603 and ISO 11801 2nd edition Category 6 standard, unscreened. To be wired according to the TIA-568 'B' convention. Colour, white.

## 13.6.3   Item 3: wall and floor outlets

- *Quantity*: 1282 dual wall sockets, single gang; 1282 shuttered modules; 968 floor boxes; 3872 'LJ6' shuttered floor box modules.

## 13.6.4   Item 4: patch panels

- *Quantity*: 273
- *Specification*: 24-port Category 6 unscreened patch panel. 19 inch, 1-U rack mount. Meeting the electrical specification of IEC 60603 Category 6 and ISO 11801 2nd edition. Cable termina-tions to be insulation displacement connectors. The unit is to be wired according to the TIA-568 'B' convention. Colour, black or grey.

## 13.6.5   Item 5: Category 6 patch cords

- *Quantity*: (a) 1500 1.5 m; (b) 1500 3.0 m.
- *Specification*: Category 6 unscreened patch cords meeting the appropriate Category 6 specification for flexible cables according to ISO 11801 2nd edition and IEC 61156 or EN 50288. Colour, grey. Sheath fire performance to meet IEC 60332-1. Each end is to be terminated with eight-position 'RJ-45' socket to IEC 60603

and ISO 11801 2nd edition Category 6 standard. The copper conductors must be flexible multicores. Each assembly must be factory made and tested and must come with guaranteed Class E/Cat 6 connectivity compliance with the Category 6 patch panels and outlets used. Note that the number of patch cords does not equal the number outlets. This is to save money in the initial purchase as not all outlets will be populated from day one. The 3-m cords will be for the work areas and the 1.5-m cords will be for the equipment areas.

## 13.6.6   Item 6: optical patch panels

- *Quantity*: 11
- *Specification*: 24-port optical patch panel, 19 inch, 1-U rack mount. The unit must be capable of terminating at least 24 single mode or multimode optical fibres with SC duplex connectors. Cables entering the back must be secured with appropriate glands. A fibre management system within the panel must be capable of organising all spare fibres and fibre splices for 1550-nm single mode operation. The unit must be on sliders to facilitate installation and repair. Colour, black or grey.

## 13.6.7   Item 7: optical adapters

- *Quantity*: (a) 120 duplex multimode; (b) 8 duplex single mode.
- *Specification*: Multimode SC duplex adapters to IEC 60874-19-3. Colour, beige. Single mode SC duplex adapters to IEC 60874-19-2. Colour, blue. To be fitted to the front of the optical patch panels as required.

## 13.6.8   Item 8: optical connectors

- *Quantity*: 240
- *Specification*: SC multimode optical connectors to IEC 60874-19-1 and IEC 61754-4. To be directly terminated onto the main fibre within the optical patch panels on-site.

### 13.6.9   Item 9: optical pigtails

- *Quantity*: 16
- *Specification*: 750–1000-mm of single mode fibre to the same specification as the main cable, supplied by the same manufacturer. Terminated at one end with SC single mode connectors to IEC 60874-14-5 and IEC 61754-4. All pigtails to be factory made and tested and individually packaged. All single mode fibres are to be terminated by fusion splicing on factory-made single mode tail cables.

### 13.6.10   Item 10: fibre optic splice protection sleeves

- *Quantity*: 16
- *Specification*: Fibre optic heat-shrink splice protection sleeves shall be applied to all fusion splices. Sleeves shall be to any major telecommunication standard, e.g. Bellcore, British Telecom, etc.

### 13.6.11   Item 11: optical patch cord assembly

- *Quantity*: (a) 120 duplex multimode; (b) 8 duplex single mode.
- *Specification*: Each optical patch cord shall be 3 m long and be constructed of tight buffered (900 µm) optical fibre to the same specification as the main cable and supplied by the same manufacturer. Patch cable to be 2.8 mm (nominal) diameter with low flammability sheath (IEC 60332-1) and containing aramid yarn strength members. Multimode patch cords are to be terminated at both ends with SC multimode duplex optical connectors to IEC 60874-19-1 and IEC 61754-4. Single mode patch cords are to be terminated at both ends with SC single mode duplex connectors to IEC 60874-14-5 and IEC 61754-4. All patch cords to be factory made and tested and individually packaged. Patch cables must be of different colours to differentiate between OM1, OM2, OM3 and OS1 optical fibres and the fibre type must be inkjet printed onto the cable sheath. Duplex patch cords shall be cross-over, i.e. A to B and B to A.

### 13.6.12    Item 12: air blown fibre multiduct

- *Quantity*: (a) four-way internal multiduct, 1500 m; (b) seven-way external multiduct, 1000 m.
- *Specification*: All fibre ducts must be dual extrusion with low friction and anti-static lining and with a low fire hazard, zero halogen sheathing to IEC 60332-1 and IEC 60754. Blown fibre ducts will have an inside diameter of 3.5 mm and an outside diameter of 5 mm. All ducts must have a working pressure of at least ten times atmospheric pressure.

    Indoor backbone multiducts will contain four individual ducts which will be oversheathed with a low fire hazard, zero halogen sheathing to IEC 60332-1 and IEC 60754.

    External grade multiducts must be oversheathed with an aluminium foil moisture barrier and black polyethylene.

    The blown fibre system must be capable of blowing fibre at least 400 m in the horizontal plane and at least 300 m in the vertical plane. The blown fibres must be able to cope with at least 300 25-mm bends within a 300 m run in any plane. The blown fibre system must be capable of transporting at least 12 individual optical fibres or one optical fibre bundle in each duct. Blown fibres must be capable of being blown out of the ducts any number of times leaving the ducting fit for re-use. Any grade of multimode or single mode fibre must be capable of being blown into the ducts.

    Ducts and multiducts pulled into conduit must have their exposed ends totally covered with a heatshrink cap or other waterproof method before installation. Cable lubricant must only be used with external grade ducts.

### 13.6.13    Item 13: air blown fibre

- *Quantity*: (a) OM3 grade, 12 000 m (to be supplied in eight reels of eight different colours); (b) OS1 grade, 8000 m (to be supplied in eight reels of eight different colours)
- *Specification*: The optical fibre specification shall be according to IEC 60793 and more specifically the multimode 50/125 fibre shall

be according to ISO 11801 2nd edition OM3 specification and the single mode shall be according to ISO 11801 2nd edition OS1 and ITU-T G652 specifications.

All fibres must be individually colour coded.

Blue    Orange
Green   Red
Yellow  Brown
Grey    Violet

All fibres must be individually coated with anti-static and low friction coating up to an outside diameter of approximately 500 μm.

## 13.6.14    Item 14: copper telephone cable

- *Quantity*: (a) internal grade, 1500 m; (b) external grade, 1000 metres
- *Specification*: 100-pair 0.5-mm copper telephone cables to ISO 11801 Category 3 specification or an appropriate national telecommunications provider specification. Internal cables will be white or cream coloured and sheathed with a low flammability material to IEC 60332-1. External cables will have an aluminium foil moisture barrier and a black polyethylene sheath. The external cables will be fully filled with petroleum gel.

## 13.6.15    Item 15: telephone cable distribution frames

- *Quantity*: 8
- *Specification*: A wall-mounted telephone cable distribution frame based on IDCs will be installed in each of the eight TRs. Each frame must accommodate up to 300 pairs. The frame installed in the campus distributor (TR G1) must also be equipped with over-voltage protectors which must be suitably earthed.

## 13.6.16    Item 16: equipment racks

- *Quantity*: 16
- *Specification*: Equipment cabinets must be 40-U high with 19 inches (IEC 297) rack mountings, with a base area of 600 mm ×

800 mm, to support the proposed equipment. The use of 600 mm × 800 mm cabinets, with the 800-mm faces to the front, is recommended as this allows for maximum installation access. Each cabinet must be supplied with base covers, fitted roof-mounted fans (when active equipment is to be installed) and fitted metal clad multiple-way switched power outlet blocks. The cabinets must have glass doors which must be lockable and offered with a common key. Normally it is recommended to utilise a glass door at the front and metal door for the rear. All glass doors must be metal framed. All cabinets must be capable of supporting rack-mounted equipment using both front and rear mountings. Support plates must be used if necessary.

### 13.6.17   Item 17: wire cable tray

- *Quantity*: 1200 m
- *Specification*: A welded steel wire cable tray coated or painted to protect it completely from corrosion. The dimensions must not be less than 500 × 72-mm. Bends must be incorporated to ensure that no cable is forced through a 90° corner. The tray must be secured to the ceiling by bolts and all accessories and hardware must be included. The cable tray must earthed and maintain earth continuity for its entire length.

## 13.7   Other issues

It is important to define what has been left out and what has been assumed in any system design. No mention has been made of the plastic trunking to be used in each office space, for example.

The assumptions made about the quality of the spaces and the rights of access and storage areas have been defined elsewhere in this book and they must be recalled in the design proposal to make it clear what is expected by the contractor.

Finally a test and handover plan must be presented as part of the documentation, defining what will be tested and how and what constitutes an acceptable pass parameter.

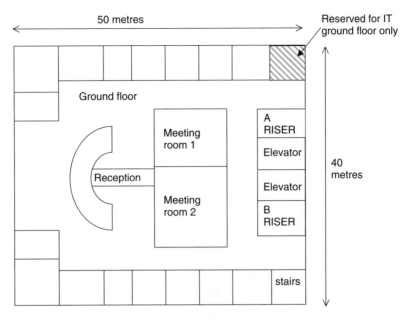

**Fig. 13.8** Ground floor of the Humphrey Laboratory.

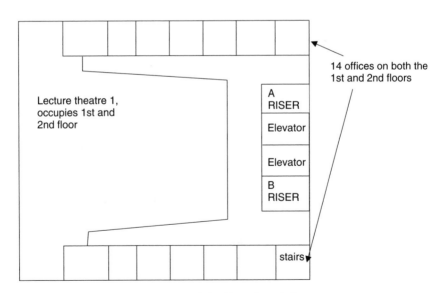

**Fig. 13.9** First and second floors of the Humphrey Laboratory.

## 13.8 The rest of the site

It is left as an exercise for the reader to finish off the site design, including the telephone cable run to the security lodge. Some clues follow for the Humphrey Laboratory as some differences will be seen when designing for the less-dense cable environment. The basic floor plans are shown in Figs. 13.8, 13.9 and 13.10.

- Note that only one TR is required. The TR nominated on the ground floor can serve the whole building.
- Allow four TOs per office and meeting room.
- Allow 12 outlets at the reception area and the lecture theatre.
- Use a density of two outlets per $10\,m^2$ metres in the laboratories.
- Use Category 6 unscreened cable.
- No optical fibre is required but remember that the single TR on the ground floor is also terminating the seven-way ABF campus multiduct and the 100-pair telephone cable.

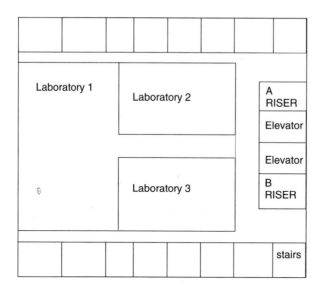

**Fig. 13.10** Third floor of the Humphrey Laboratory.

# Appendix I: List of relevant standards

It would be impossible to assemble every relevant standard in the world but the majority of European and international standards are listed here along with the most relevant American (USA) ones. Standards are always changing and readers should check for latest editions from the publishers, who are listed in Appendix II.

| | |
|---|---|
| af-phy-0015.000 | ATM Forum. Physical Medium Dependent Interface Specification for 155 Mbps over twisted pair cable. |
| af-phy-0018.000 | ATM Forum. Mid range physical layer specification for Category 3 UTP. |
| af-phy-0040.000 | ATM Forum. Physical interface specification for 25.6 Mbps over twisted pair. |
| af-phy-0046.000 | ATM Forum. 622.08 Mbps physical layer. |
| af-phy-0047.000 | ATM Forum. 155.52 Mbps Physical layer specification for Category 3 UTP. |
| af-phy-0053.000 | ATM Forum. 120-ohm addendum for 155 Mbps over twisted pair. |
| af-phy-0062.000 | ATM Forum. 155 Mbps over MMF short wave length lasers. |
| af-phy-0079.000 | ATM Forum. 155 Mbps over plastic optical fibre. |
| af-phy-0079.001 | ATM Forum. 155 Mbps over Hard Clad polymer fibre. |

|  |  |
|---|---|
| af-phy-0110.000 | ATM Forum. Physical layer high density glass optical fibre annex. |
| af-phy-0133.000 | ATM Forum. 2.4 Gbps Physical layer specification. |
| af-phy-0162.000 | ATM Forum. 1000 mb/s over fibre and category 6. |

*Note: look for TIA standards under 'TIA' as well as 'ANSI'*

|  |  |
|---|---|
| ANSI X3T9.3/91-005 | HIPPI Physical layer. |
| ANSI/EIA/TIA-492AAAA-A-1998 | Detail specification for 62.5/125 Class 1a multimode, graded index, optical waveguide fibers. |
| ANSI/EIA/TIA-492AAAB | Detail specification for 50/125 class 1a multimode, graded index, optical waveguide fibers. |
| ANSI/EIA/TIA-492CAAA-1998 | Detail specification for class IVa dispersion-unshifted singlemode, optical fibers. |
| ANSI/ICEA S-83-596-1994 | Fiber Optic Premises Distribution Cable. |
| ANSI/ICEA-S-87-640-2000 | Fiber Optic Outside Plant Communications Cable. |
| ANSI/NECA/BICSI 568-2001 | Installing Commercial Building Telecommunications Cabling. |
| ANSI/TIA/EIA-526-14-A | Optical Power Loss Measurements of Installed Multimode Fiber Cable Plant. |
| ANSI/TIA/EIA-526-7 | Optical Power Loss Measurements of Installed Singlemode Fiber Cable Plant. |
| ANSI/TIA/EIA-568-B | Commercial Building Telecommunications Cabling Standard – 2002. |

Note: ANSI/TIA/EIA-568-A was replaced by 568-B in 2001, but it is listed here for the completeness of the reference

|  |  |
|---|---|
| TIA/EIA 568-A | Commercial Building Telecommunications Cabling Standard, 1995. |
| Addendum 1 – | Propagation Delay and Delay Skew, Sept 1997. |
| Addendum 2 – | Corrections and Additions to TIA-568-A, Aug 1998. |

| | |
|---|---|
| Addendum 3 – | Hybrid Cables, Dec 1998. |
| Addendum 4 – | Patch Cord Qualification Test, Aug 1999. |
| Addendum 5 – | Additional Transmission Performance Specifications for 4-Pair 100 ohm Category 5e Cabling Nov 1999. |
| TIA/EIA TSB 67 | Telecommunications Systems Bulletin. Transmission Performance Specifications for Field Testing of Unshielded Twisted Pair Cabling Systems, Oct 1995. |
| TIA/EIA TSB 72 | Telecommunications Systems Bulletin Centralized Optical Fiber Cabling Guidelines. Oct 1995. |
| TIA/EIA TSB 75 | Telecommunications Systems Bulletin Additional Horizontal Cabling Practices for Open Offices Aug 1996. |
| TIA/EIA TSB 95 | Telecommunications Systems Bulletin Additional Transmission Performance Guidelines for 100 Ohm 4 pair Category 5 Cabling. August 1999. |
| ANSI/TIA/EIA-568-B.1-2 | Grounding and Bonding Specifications for Screened Horizontal Cabling. |
| ANSI/TIA/EIA-568-B.3.1 | Additional Transmission Performance Specifications for 50/125 Optical Fiber Cables. |
| ANSI/TIA/EIA-568-B.2.1 | Transmission Performance Specifications for 4-pair Category 6 Cabling. |
| ANSI/TIA/EIA-569-A | Commercial Building Standard for Telecommunications Pathways and Spaces. |
| ANSI/TIA/EIA-570-A | Residential Telecommunications Cabling Standard. |
| ANSI/TIA/EIA-606 | Administration Standard for the Telecommunications Infrastructure of Commercial Buildings. |
| ANSI/TIA/EIA-607 | Commercial Building Grounding and Bonding Requirements for Telecommunications. |

| | |
|---|---|
| AS/NZS 3080 | Telecommunications installations – Integrated telecommunications cabling systems for commercial premises, Australia and New Zealand. |
| BS 6701 | Code of practice for installation of apparatus intended for connection to certain telecommunications systems. |
| BS 7671 | Requirements for Electrical Installations, also known as the IEE Wiring Regulations 16th Edition and its corresponding Guidance Note No. 5 Protection Against Electric Shock. |
| BS 7718 | Code of Practice for Fibre Optic Cabling. |
| CAN/CSA T529 | Commercial building telecommunications cabling standard – Canada. |
| CISPR 22 | Limits and methods of measurement of radio disturbance characteristics of information technology equipment. |
| CISPR 24 | Information technology equipment – Immunity characteristics – Limits and methods of measurement. |
| DISC PD1002 | A Guide to Cabling in Private Telecommunications Systems (TIA UK). |
| EN 187 000 | Generic specification for optical fibre cables. |
| EN 188 000 | Generic specification for optical fibres. |
| EN 28877 | (ISO/IEC 8877) Information Technology. Telecommunications and information exchange between systems. Interface connector and contact assignments for ISDN Basic Access interface located at reference points S and T. |
| EN 41003 | 1998 Particular safety requirements for equipment to be connected to telecommunication networks. |

| | |
|---|---|
| EN 50082-1 | Electromagnetic compatibility – Generic immunity standard – Part 1: Residential, commercial and light industry. |
| EN 50082-2 | Electromagnetic compatibility – Generic immunity standard – Part 2: Industrial environment. |
| EN 50085 | Cable trunking systems and cable ducting systems for electrical installations. |
| EN 50086 | Conduit systems for electrical installations requirements. |
| EN 50098-1 | Customer premises cabling for information technology – Part 1: ISDN basic access. |
| EN 50098-2 | Customer premises cabling for information technology – Part 2: 2048 kb/s ISDN primary access and leased line network interface. |
| EN 50130 | Alarm systems – Part 4. Electromagnetic compatibility – Product family standard: Immunity requirements for components of fire, intruder and social alarm systems. |
| EN 50167 | Sectional specifications for horizontal floor wiring cables with a common overall screen for use in digital communications. |
| EN 50168 | Sectional specifications for work-area wiring cables with a common overall screen for use in digital communications. |
| EN 50169 | Sectional specifications for backbone cables, riser and campus with a common overall screen for use in digital communications. |
| EN 50173 | Information technology, generic cabling systems, August 1995, with Amendment No. 1 published in January 2000. |
| EN 50174-1 | Information technology – Cabling installation – Part 1: Specification and quality assurance. |

| EN 50174-2 | Information technology – Cabling installation – Part 2: Installation and planning practices inside buildings. |
| EN 50174-3 | Information technology – Cabling installation – Part 3: Installation and planning practices external to buildings. |
| EN 50265-2-1 | (IEC 60332-1): Flammability, single cable. |
| EN 50266-2-4 | (IEC 60332-3): Flammability of a bunch of cables. |
| EN 50267 | (IEC 60754): Acidity and conductivity. |
| EN 50288 | Multi-element metallic cables used in analogue and digital communications and control. (Note: replaces EN 50167, EN 50168, and EN 50169). |
| EN 50288-2-1 | 100 MHz Screened, horizontal and backbone. |
| EN 50288-2-2 | 100 MHz Screened, patch. |
| EN 50288-3-1 | 100 MHz Unscreened, horizontal and backbone. |
| EN 50288-3-2 | 100 MHz Unscreened, patch. |
| EN 50288-4-1 | 600 MHz Screened, horizontal and backbone. |
| EN 50288-4-2 | 600 MHz Screened, patch. |
| EN 50288-5-1 | 200 MHz Screened, horizontal and backbone. |
| EN 50288-5-2 | 200 MHz Screened, patch. |
| EN 50288-6-1 | 200 MHz Unscreened, horizontal and backbone. |
| EN 50288-6-2 | 200 MHz Unscreened, patch. |
| EN 50289 | Communication cables – Specification for test methods. |
| EN 50289-1-15 | Communication cables – Specifications for test methods – Part 1–15: Coupling attenuation. |
| EN 50289-4-11 | Flame propagation, heat release, time to ignition, flaming droplets. |

| | |
|---|---|
| EN 50310 | Application of equipotential bonding and earthing in buildings with information technology equipment. |
| EN 50346 | Information technology – Cabling installation – Testing of installed cabling. |
| EN 50368 | (IEC 61034) Smoke evolution. |
| EN 55013 | Limits and methods of measurement of radio disturbance characteristics of broadcast receivers and associated equipment. |
| EN 55020 | Electromagnetic immunity of broadcast receivers and associated equipment. |
| EN 55024 | Information technology equipment. immunity characteristics. Limits and methods of measurement. |
| EN 55103 | Electromagnetic compatibility – Product family standard for audio, video, audio-visual and entertainment lighting control system for professional use. |
| EN 60825 | Safety of laser products (IEC 60825). |
| EN 60825-1 | Safety of laser products – Part 1: Equipment classification, requirements and user's guide (IEC 60825-1:1993). |
| EN 60950 | Safety of information technology equipment (IEC 60950). |
| EN 61000 | (IEC 61000) Electromagnetic compatibility environment (contains over 17 parts). |
| EN 61280-4-2 | Fibre optic communication subsystem basic test procedures – Part 4–2: Fibre optic cable plant – Single-mode fibre optic cable plant attenuation (IEC 61280-4-2:1999). |
| EN 61300-3-34 | Fibre optic interconnecting devices and passive components – Basic test and measurement procedures – Part 3–34: |

|  | Examinations and measurements – Attenuation of random mated connectors (IEC 61300-3-34:1997). |
| EN 61300-3-6 | Fibre optic interconnecting devices and passive components – Basic test and measurement procedures – Part 3–6: Examinations and measurements – Return loss (IEC 61300-3-6:1997). |
| EN 61537 | Cable tray and cable ladder systems for electrical installations. |
| EN 61935-1 | Generic cabling systems – Specification for the testing of balanced communications cabling in accordance with EN 50173 – Part 1: Installed cabling (IEC 61935-1:2000). |
| ETS 300 253 | Equipment engineering (EE) – Earthing and bonding of telecommunication equipment in telecommunication centres. |
| HD 384.2 S1 | 1986 International electrotechnical vocabulary – Chapter 826: Electrical installations of buildings (IEC 60050-826:1982). |
| HD 384.3 S2 | 1995 Electrical installations of buildings – Part 3: Assessment of general characteristics of installations (IEC 60364-3:1993, modified). |
| HD 384.4.41 S2 | 1996 Electrical installations of buildings – Part 4: Protection for safety – Chapter 41: Protection against electric shock (IEC 60364-4-41:1992, modified). |
| HD 384.5.54 S1 | 1988 Electrical installations of buildings – Part 5: Selection and erection of electrical equipment – Chapter 54: Earthing arrangements and protective conductors (IEC 60364-5-54:1980, modified). |
| ICEA S-100-685 | Station wire for indoor/outdoor use. |

| | |
|---|---|
| ICEA S-101-699 | Cat 3 station wire and inside wiring cables up to 600 pairs. |
| ICEA S-102-700 | Cat 5, 4-pair, indoor UTP wiring standard. |
| ICEA S-103-701 | AR&M riser cable. |
| ICEA S-80-576 | Communications wire and cable for wiring of premises. |
| ICEA S-83-596 | Optical fiber indoor/outdoor cable. |
| ICEA S-87-640 | Optical fiber outside plant cable. |
| ICEA S-89-648 | Aerial service wire. |
| ICEA S-90-661 | Individually unshielded twisted pair indoor cables. |

Note: check for IEC standards under 'ISO' as well.

| | |
|---|---|
| IEC 61034 | Smoke density and evolution. |
| IEC 60050-604 | 1987 International electrotechnical vocabulary – Chapter 604: Generation, transmission and distribution of electricity – Operation. |
| IEC 60068-1 | 1988 Environmental testing – Part 1: General and guidance. |
| IEC 60068-2-14 | Environmental testing – Part 2: Tests – Test N: Change of temperature. |
| IEC 60068-2-2 | Basic environmental testing procedures – Part 2: Tests – Tests B: Dry heat. |
| IEC 60068-2-38 | Environmental testing – Part 2: Tests – Test Z/AD: Composite temperature/humidity cyclic test. |
| IEC 60068-2-6 | Environmental testing – Part 2: Tests – Tests Fc: Vibration (sinusoidal). |
| IEC 60068-2-60 | Environmental testing – Part 2: Tests – Test Ke: Flowing mixed gas corrosion test. |
| IEC 60096-1 | Radio-frequency cables – Part 1: General requirements and measuring methods. |
| IEC 60189-1 | Low-frequency cables and wires with p.v.c. insulation and p.v.c. sheath – Part 1: General test and measuring methods. |

| | |
|---|---|
| IEC 60227-2 | Polyvinyl chloride insulated cables of rated voltages up to and including 450/750 V – Part 2: Test methods. |
| IEC 60300 | Series: Dependability management. |
| IEC 60332-1 | Flammability of a single vertical cable. |
| IEC 60332-3 | Test on electric cables under fire conditions – Part 3: Tests on bunched wires or cables. |
| IEC 60332-3-24c | Flammability of a bunch of vertical cables. |
| IEC 60332-3-c | Flammability of a bunch of vertical cables. |
| IEC 60352-3 | 1993: Solderless connections – Part 3: Solderless accessible insulation displacement connections – General requirements, test methods and practical guidance. |
| IEC 60352-4 | 1994: Solderless connections – Part 4: Solderless non-accessible insulation displacement connections – General requirements, test methods and practical guidance. |
| IEC 60352-6 | Solderless connections – Part 6: Insulation piercing connections – General requirements, test methods and practical guidance. |
| IEC 60364 | Electrical installations of buildings |
| IEC 60364-1 | Electrical installation of buildings – Part 1: Fundamental principles, assessment of general characteristics, definitions. |
| IEC 60364-4-41 | Electrical installation of buildings – Part 4–41: Protection for safety – Protection against electric shock. |
| IEC 60364-5-548 | Electrical installation of buildings – Part 5: Selection and erection of electrical equipment – Section 548: Earthing arrangements and equipotential bonding for information technology installations. |
| IEC 60512 | Series, Electromechanical components for electronic equipment. |

| IEC 60512-1 | Connectors for electronic equipment – Tests and measurements – Part 1: General. |
| IEC 60512-2 | 1985: Electromechanical components for electronic equipment; basic testing procedures and measuring methods – Part 2: General examination, electrical continuity and contact resistance tests, insulation tests and voltage stress tests Amendment 1 (1988). |
| IEC 60512-25-1 | RDIS: Electromechanical components for electronic equipment – Basic testing procedures and measuring methods – Part 25–1: Test 25a Crosstalk ratio test procedure for electrical connectors, sockets and cable assemblies. |
| IEC 60512-25-2 | ADIS: Electromechanical components for electronic equipment – Basic testing procedures and measuring methods – Part 25–2: High speed electronics tests – Test 25b Attenuation/insertion loss. |
| IEC 60512-25-3 | RDIS: Electromechanical components for electronic equipment – Basic testing procedures and measuring methods – Part 25–3: Test 25c Rise time degradation test procedure. |
| IEC 60512-25-4 | RDIS: Electromechanical components for electronic equipment – Basic testing procedures and measuring methods – Part 25–4: Test 25d Propagation delay test procedure for electrical connectors, sockets, cable assemblies or interconnection systems. |
| IEC 60512-25-5 | CCDV: Connectors for electronic equipment – Basic tests and measurements – Part 25–5: Test 25e Return loss. |
| IEC 60512-3 | 1976: Electromechanical components for electronic equipment; Basic testing |

| | |
|---|---|
| | procedures and measuring methods. Part 3: Current-carrying capacity tests. |
| IEC 60603-7 | Amendment 1. Detail specification for connectors 8 way. Test methods and related requirements for use at frequencies up to 100 MHz. |
| IEC 60603-7 | 1996: Connectors for frequencies below 3 MHz for use with printed boards – Part 7: Detail specification for connectors, 8 way, including fixed and free connectors with common mating features. |
| IEC 60603-7-1 | ADIS: Inclusion of screen mating in IEC 60603-7. |
| IEC 60603-7-2 | ADIS: Detail specification for 8 way unshielded connectors, with assessed quality, including fixed and free connectors with common mounting features; test methods and related requirements for use at frequencies up to 100 MHz. |
| IEC 60603-7-3 | ADIS: Detail specification for 8 way shielded connectors, with assessed quality, including fixed and free connectors with common mounting features; test methods and related requirements for use at frequencies up to 100 MHz. |
| IEC 60603-7-4 | CD: Connectors for electronic equipment: Detail specification for an unshielded 8 way connector with performance up to 250 MHz. |
| IEC 60603-7-5 | ADIS: Detail specification for 8 way shielded connectors, with assessed quality, including fixed and free connectors with common mounting features; test methods and related requirements for use at frequencies up to 250 MHz. |
| IEC 60603-7-7 | CCDV: Connectors for use in DC, low frequency analogue and in digital high |

| | |
|---|---|
| | speed data applications – Part 7–7: 8 way connectors for frequencies up to 600 MHz [Category 7 Detail Specification]. |
| IEC 60708-1 | Low-frequency cables with polyolefin insulation and moisture barrier polyolefin sheath – Part 1: General design details and requirements. |
| IEC 60754-1 | Halogen gas emission. |
| IEC 60754-2 | Smoke corrosivity. |
| IEC 60793-1-1 (1999-02) Ed. 1.1 | Consolidated edition optical fibres – Part 1–1: Generic specification – General. |
| IEC 60793-1-4 | Optical fibres – Part 1: Generic specification – Section 4: Measuring methods for transmission and optical characteristics. |
| IEC 60793-1-40 | Optical fibres – Part 1–40: Measurement methods and test procedures – Attenuation. |
| IEC 60793-1-41 | Optical fibres – Part 1–41: Measurement methods and test procedures – Bandwidth. |
| IEC 60793-1-44 | Optical fibres – Part 1–44: Measurement methods and test procedures – Cut-off wavelength. |
| IEC 60793-2 | 1992: Optical fibres – Part 2: Product specifications. |
| IEC 60793-2-10 Ed. 1.0 | Optical Fibres – Part 2–10: Product specifications – Sectional specification for category A1 multimode fibres. |
| IEC 60793-2-20 (2001-12) | Optical fibres – Part 2–20: Product specifications – Sectional specification for category A2 multimode fibres. |
| IEC 60793-2-30 Ed. 1.0 | Optical fibres – Part 2–30: Product specifications – Sectional specification for category A3 multimode fibres. |
| IEC 60793-2-40 Ed. 1.0 | Optical fibres – Part 2–40: Product specifications – Sectional specification for category A4 multimode fibres. |

| | |
|---|---|
| IEC 60793-2-50 Ed. 1.0 | Optical fibres – Part 2–50: Product specifications – Sectional specification for class B single-mode fibres. |
| IEC 60794-1-1 (2001-07) | Optical fibre cables – Part 1–1: Generic specification. |
| IEC 60794-2 (1998-08) Ed. 2.1 | Consolidated edition optical fibre cables – Part 2: Product specification (indoor cable). |
| IEC 60794-1-1 | Optical fibre cables – Part 1–1: Generic specification – General. |
| IEC 60794-1-2 | Optical fibre cables – Part 1–2: Generic specification – Basic optical cable test procedures. |
| IEC 60794-2 (1998-08) Ed. 2.1 | Consolidated Edition Optical fibre cables – Part 2: Product specification (indoor cable). |
| IEC 60794-2-10 | *Simplex and duplex cables.* |
| IEC 60794-2-20 | *Multi-fibre cables.* |
| IEC 60794-2-30 | Ribbon cords. |
| IEC 60794-3 (2001-09) | Optical fibre cables – Part 3: Sectional specification – Outdoor cables. |
| IEC 60794-3-10 | *Duct or buried cables.* |
| IEC 60794-3-20 | *Aerial cables.* |
| IEC 60794-3-30 | *Underwater cables.* |
| IEC 60794-4-1 (1999-01) | Optical fibre cables – Part 4–1: Aerial optical cables for high-voltage power lines. |
| IEC 60801 | Electrostatic discharge and electrical fast transient immunity. |
| IEC 60807-8 | Rectangular connectors for frequencies below 3 MHz – Part 8: Detailed specification for connectors, four signal contacts and earthing contacts for cable screen. |
| IEC 60811-1-1 | Common test methods for insulating and sheathing materials of electric cables – Part 1: Methods for general application – Section 1: Measurement of thickness and overall dimensions – Tests for determining the mechanical properties. |

| | |
|---|---|
| IEC 60825-1 | Safety of laser products – Part 1: Equipment classification, requirements and user's guide. |
| IEC 60874-1 | Connectors for optical fibres and cables – Part 1: Generic specification. |
| IEC 60874-7 | 1993: FC optical connector. |
| IEC 60874-10 | Connectors for optical fibres and cables – Part 10: Sectional specification for fibre optic connector – Type BFOC/2,5. |
| IEC 60874-10-1 | 1997: BFOC/2.5 multimode. |
| IEC 60874-10-2 | 1997: BFOC/2.5 single mode. |
| IEC 60874-10-3 | 1997: BFOC/2.5 single mode and multimode. |
| IEC 60874-14 | 1993: SC. |
| IEC 60874-14-1 | 1997: SC/PC multimode. |
| IEC 60874-14-2 | 1997: SC/PC tuned single mode. |
| IEC 60874-14-3 | 1997: SC single mode simplex adaptor. |
| IEC 60874-14-4 | 1997: SC multimode simplex adaptor. |
| IEC 60874-14-5 | 1997: SC/PC untuned single mode. |
| IEC 60874-14-6 | 1997: SC APC 9° untuned single mode. |
| IEC 60874-14-7 | 1997: SC APC 9° tuned single mode. |
| IEC 60874-14-9 | 1999: SC APC 8° tuned single mode. |
| IEC 60874-14-10 | 1999: SC APC 8° untuned single mode. |
| IEC 60874-16 | 1994: MT. |
| IEC 60874-19 | (all parts), Connectors for optical fibres and cables. |
| IEC 60874-19-1 | Ed. 1.0 PPUB: Connectors for optical fibres and cables – Part 19–1: Fibre optic patch cord connector type SC-PC (floating duplex) standard terminated on multimode optical fibre type A1a, A1b – Detail specification. |
| IEC 60874-19-2 | Ed. 1.0 PPUB: Connectors for optical fibres and cables – Part 19–2: Fibre optic adaptor (duplex) type SC for single mode fibre connectors – Detail specification. |

| | |
|---|---|
| IEC 60874-19-3 | Ed. 1.0 PPUB: Connectors for optical fibres and cables – Part 19–3: Fibre optic adaptor (duplex) type SC for multimode fibre connectors – Detail specification. |
| IEC 60874-19-4 | Detail specification for fibre optic connector (duplex) type SC-PC premium for multimode fibre type A1a, A1b. |
| IEC 60874-19-5 | Detail specification for fibre optic connector Type SC-PC (rigid duplex) standard for multimode fibre type A1a, A1b. |
| IEC 60950 | Safety of information technology equipment, including electrical business equipment. |
| IEC 61000-2-2 | Electromagnetic compatibility (EMC) – Part 2: Environment – Section 2: Compatibility levels for low-frequency conductor disturbances and signalling in public low-voltage power supply systems. |
| IEC 61000-5-2 | Electromagnetic compatibility (EMC) – Part 5: Installation and mitigation. |
| IEC 61024 | Protection of structures against lightning. |
| IEC 61034 | Smoke evolution. |
| IEC 61035-1 | Specification for conduit fittings for electrical installations – Part 1: General requirements. |
| IEC 61073-1 | 1994 Splices for optical fibres and cables – Part 1: Generic specification – Hardware and accessories. |
| IEC 61076-3-104 | (under consideration) Connectors with assessed quality for use in DC, low frequency analogue and in digital high speed data applications – Part 3–104: 8–way connectors for frequencies up to 600 MHz [Category 7 Detail Specification]. |
| IEC 61140 | Protection against electric shock. |
| IEC 61156 | Multicore and symmetrical pair/quad cables for digital communications. |

| | |
|---|---|
| IEC 61156-1 | Multicore and symmetrical pair/quad cables for digital communications – Part 1: generic specification. |
| IEC 61156-2 | Multicore and symmetrical pair/quad cables for digital communications – Part 2: Horizontal floor wiring – Sectional specification. |
| IEC 61156-3 | Multicore and symmetrical pair/quad cables for digital communications – Part 3: Work area wiring – Sectional specification. |
| IEC 61156-5 | CD: Multicore and symmetrical pair/quad cables for digital communication – Part 5: Symmetrical pair/quad cables with transmission characteristics up to 600 MHz – Sectional specification. |
| IEC 61156-6 | CCDV: Symmetrical pair/quad cables for digital communications with transmission characteristics up to 600 MHz – Part 6: Work area wiring – Sectional specification. |
| IEC 61280 | Series: Fibre optic communication subsystem basic test procedures. |
| IEC 61280-1-1 | Fibre optic communication subsystem basic test procedures – Part 1.1: Test procedures for general communications subsystems – Transmitter output optical power measurement for single mode optical fibre cable. |
| IEC 61280-4-1 | Fibre optic communication subsystem basic test procedures – Part 4.1: Test procedures for fibre optic cable plant – Multimode fibre optic plant attenuation measurement. |
| IEC 61280-4-2 | Fibre optic communication subsystem basic test procedures – Part 4.2: Test procedures for fibre optic cable plant – Single mode fibre optic plant attenuation measurement. |

| | |
|---|---|
| IEC 61280-4-3 | Fibre optic communication subsystem basic test procedures – Part 4.3: Test procedures for fibre optic cable plant – Single mode fibre optic plant optical return loss measurement. |
| IEC/TR 61282-1 | Ed. 1.0 Fibre optic communication system design guides – Part 1: Single mode digital and analogue systems. |
| IEC 61282-2 TR | Fibre optic communication system design guides – Part 2: Multimode and single mode Gb/s applications – Gigabit Ethernet model. |
| IEC 61282-3 TR | Ed. 1.0 Fibre optic communication system design guides: Calculation of PMD in fibre optic systems. |
| IEC 61282-4 TR | Ed. 1.0 Fibre optic communication system design guides – Part 4: Guideline to accommodate and utilise non-linear effects in single-mode fibre optic systems. |
| IEC/TR 61282-5 | Ed. 1.0 Fibre optic communication system design guides – Part 5: Accommodation and compensation of dispersion. |
| IEC 61282-6 TR | Ed. 1.0 E Fibre optic communication system design guides – Part 6: Skew design in parallel optical interconnection systems. |
| IEC 61282-7 TR | Ed. 1.0 Fibre optic communication system design guides – Part 7: Statistical calculation of chromatic dispersion. |
| IEC 61300-2-1 | Fibre optic interconnecting devices and passive components – Basic test and measurement procedures – Part 2–1: Tests – Vibration (sinusoidal) – Edition 2. |
| IEC 61300-2-2 | Fibre optic interconnecting devices and passive components – Basic test and measurement procedures – Part 2–2: Tests – Mating durability. |

| | |
|---|---|
| IEC 61300-2-4 | Fibre optic interconnecting devices and passive components – Basic test and measurement procedures – Part 2–4: Tests – Fibre/cable retention. |
| IEC 61300-2-5 | Fibre optic interconnecting devices and passive components – Basic test and measurement procedures – Part 2–5: Tests – Torsion/twist. |
| IEC 61300-2-6 | Fibre optic interconnecting devices and passive components – Basic test and measurement procedures – Part 2–6: Tests – Tensile strength of coupling mechanism. |
| IEC 61300-2-12 | Fibre optic interconnecting devices and passive components – Basic test and measurement procedures – Part 2–12: Tests – Impact. |
| IEC 61300-2-17 | Fibre optic interconnecting devices and passive components – Basic test and measurement procedures – Part 2–17: Tests – Cold. |
| IEC 61300-2-18 | Fibre optic interconnecting devices and passive components – Basic test and measurement procedures – Part 2–18: Tests – Dry heat – High temperature endurance. |
| IEC 61300-2-19 | Fibre optic interconnecting devices and passive components – Basic test and measurement procedures – Part 2–19: Tests – Damp heat (steady state). |
| IEC 61300-2-22 | Fibre optic interconnecting devices and passive components – Basic test and measurement procedures – Part 2–22: Tests – Change of temperature. |
| IEC 61300-2-42 | Fibre optic interconnecting devices and passive components – Basic test and measurement procedures – Part 2–42: Tests – Static side load for connectors. |

| | |
|---|---|
| IEC 61300-3-6 | Fibre optic interconnecting devices and passive components – Basic test and measurement procedures – Part 3–6: Examinations and measurements – Return loss. |
| IEC 61312 | Protection against lightning electromagnetic impulse. |
| IEC 61753-1-1 | Fibre optic interconnecting devices and passive components performance standard – Part 1–1: General and guidance – Interconnecting devices (connectors). |
| IEC 61754-2 | 1996: BFOC/2.5 (ST). |
| IEC 61754-4 | 2000: SC. |
| IEC 61754-5 | 1996: MT. |
| IEC 61754-6 | 1997: MU. |
| IEC 61754-7 | 2000: MPO. |
| IEC 61754-13 | 1999: FC-PC. |
| IEC 61754-18 | draft: MT-RJ. |
| IEC 61754-19 | draft: SG. |
| IEC 61754-20 | draft: LC. |
| IEC 61935-1 | Generic cabling systems – Specification for the testing of balanced communication cabling in accordance with ISO/IEC 11801 – Part 1: Installed cabling. |
| IEC 61935-2 | CDV: Generic cabling systems – Specification for the testing of balanced communication cabling in accordance with ISO/IEC 11801 – Part 2: Patchcord and work area cabling. |
| IEC/TR3 61000-5-2 | Electromagnetic compatibility (EMC) – Part 5: Installation and mitigation guidelines – Section 2: Earthing and cabling. |
| IEEE 802.3ab | Physical layer specification for 1000 Mb/s operation on four pairs of Category 5 or better balanced twisted pair cable (1000BaseT), July 1999. |
| IEEE 802.3z | Media access control (MAC) parameters, physical layer, repeater and management |

parameters for 1000 Mb/s operation. June 1998.

*Note: check for ISO standards under 'IEC' as well.*

| | |
|---|---|
| ISO 11801 2nd edition | Information technology – Cabling for customer premises 2002: ISO/IEC 11801 Ed.1.2: 2000 ISO/IEC 11801 Ed.1: 1995 |
| ISO 15018 | Integrated cabling for all services other than mains power in homes, SoHo and buildings. |
| ISO 18010 | (draft) Information technology: pathways and spaces for customer premises cabling. |
| ISO/IEC 11518-1 | Information technology – High performance parallel interface – Part 1: Mechanical, electrical and signalling protocol specification (HIPPI-PH). |
| ISO/IEC 14165-1 | Information technology – Fibre channel – Part 1: Physical and Signaling Interface. |
| ISO/IEC 14709-1 | Information technology – Configuration of customer premises cabling (CPC) for applications – Part 1: Integrated services digital network (ISDN) basic access. |
| ISO/IEC 14709-2 | Information technology – Configuration of customer premises cabling (CPC) for applications – Part 2: Integrated services digital network (ISDN) primary rate. |
| ISO/IEC 14763-1 | Information technology – Implementation and operation of customer premises cabling – Part 1: Administration. |
| ISO/IEC 14763-2 TRT3 | Information technology – Implementation and operation of customer premises cabling – Part 2: Planning and installation. |
| ISO/IEC 14763-3 TRT3 | Information technology – Implementation and operation of customer premises cabling – Part 3: Testing of optical fibre cabling. |

| | |
|---|---|
| ISO/IEC 8802-12 | Information technology – Local and metropolitan area networks specific requirements – Part 12: Demand-priority access method, physical layer and repeater specifications. |
| ISO/IEC 8802-3 | Information technology – Local and metropolitan area networks – Part 3: Carrier sense multiple access with collision detection (CSMA/CD) access method and physical layer specifications. |
| ISO/IEC 8802-4 | Information processing systems – Local area networks – Part 4: Token-passing bus access method and physical layer specifications. |
| ISO/IEC 8802-5 | Information technology – Local and metropolitan area networks – Part 5: Token ring access method and physical layer specifications. |
| ISO/IEC 8802-9 | Information technology – Local and metropolitan area networks specific requirements – Part 9: Integrated services (IS) LAN interface at the medium access control (MAC) and physical (PHY) layers. |
| ISO/IEC 8877 | Information technology – Telecommunications and information exchange between systems – Interface connector and contact assignment for ISDN basic access interface located at reference points S and T. |
| ISO/IEC 9314-10 | AFCD Information technology – Fibre distributed data interface (FDDI) – Part 10: Twisted pair physical layer medium dependent (TP-PMD). |
| ISO/IEC 9314-3 | Information processing systems – Fibre distributed data interface (FDDI) – Part 3: Physical layer medium dependent (PMD). |
| ISO/IEC 9314-4 | Information processing systems – Fibre |

|  | distributed data interface (FDDI) – Part 4: Single mode fibre physical layer medium dependent (SMF-PMD). |
| ISO/IEC 9314-9 | Information processing systems; fibre distributed data interface (FDDI) – Part 9: Low cost fibre, physical layer medium dependent (LCF-PMD). |
| ISO/IEC ADIS 14165-111 | Information technology – Fibre channel, physical and signalling interface. |
| ISO/IEC/TR 12075 | Information technology – Customer premises cabling – Planning and installation guide to support ISO/IEC 8802-5 token ring stations. |
| ISO/IEC/TR2 11802-4 | 1994: Information technology – Telecommunications and information exchange between systems – Local and metropolitan area networks – Technical reports and guidelines – Part 4: Token ring access method and physical layer specifications – Fibre optic attachment. |
| ITU-T Handbook | Earthing of telecommunication installations (Geneva 1976). |
| ITU-T K.27 | 1996: Bonding configurations and earthing inside a telecommunication building. |
| ITU-T K.31 | 1993: Bonding configurations and earthing of telecommunication installations inside a subscriber's building. |
| ITU-T Rec. G.117 | Transmission aspects of unbalance about earth. |
| ITU-T Rec. G.650 | Transmission media characteristics – Definition and test methods for the relevant parameters of single mode fibres. |
| ITU-T Rec. G.651 | Characteristics of a 50/125 μm multimode graded index optical fibre cable. |
| ITU-T Rec. G.652 | 1993: Characteristics of a single mode optical fibre cable. |

| | |
|---|---|
| ITU-T G.653 | Characteristics of a dispersion-shifted single mode optical fibre cable. |
| ITU-T G.654 | Characteristics of a cut-off shifted single mode optical fibre cable. |
| ITU-T G.655 | Characteristics of a non-zero dispersion shifted single mode optical fibre cable. |
| ITU-T Rec. I.430 | Basic user-network interface; Layer 1 specification. |
| ITU-T Rec. I.431 | Primary rate user-network interface; Layer 1 specification. |
| ITU-T Rec. I.432 | B-ISDN user network interface; Physical layer specification. |
| ITU-T Rec. O.9 | Measuring arrangements to assess the degree of unbalance about earth. |
| ITU-T Rec. V.11 | Electrical characteristics for balanced double-current interchange circuits for general use with integrated circuit equipment in the field of data communications. |
| ITU-T Rec. X.21 | Interface between data terminal equipment (DTE) and data circuit-terminating equipment (DCE) for synchronous operation on public data networks. |
| NEMA WC-63.1 | Performance standards for twisted pair premise voice and data communications cable. |
| NEMA WC-63.2 | Performance standards for coaxial communications cable. |
| NEMA WC-66.1 | Performance standards for Cat 6, Cat 7 100-ohm shielded and unshielded twisted pair cables. |

*Note: Check for TIA and EIA standards under 'ANSI' as well.*

| | |
|---|---|
| TIA/EIA 604-1 | April 1996: Biconic. |
| TIA/EIA 604-2 | November 1997: ST. |
| TIA/EIA 604-3 | August 1997: SC. |
| TIA/EIA 604-4 | August 1997: FC. |

| | |
|---|---|
| TIA/EIA 604-5 | November 1999: MPO. |
| TIA/EIA 604-6 | March 1999: Fiber jack. |
| TIA/EIA 604-7 | January 1999: SG (VF-45). |
| TIA/EIA 604-10 | October 1999: LC. |
| TIA/EIA 604-12 | September 2000: MT-RJ. |
| TIA/EIA-785 | 100 mb/s Physical Layer Medium Dependant Sublayer and 10 mb/s Auto-negotiation on 850 nm Fiber optics |
| TIA/EIA-854 | A full Duplex Ethernet Physical Layer Specification for 1000 mbits/s operating over Category 6 Balanced Twisting Pair Cabling 1000 BASE-TX |
| UL 1581 | Reference standard for electrical wires, cables and flexible cords (i.e. general purpose). |
| UL 1666 | Test for flame propagation height of electrical and optical fiber cables installed vertically in shafts (i.e. riser). |
| UL 910 | Test for flame propagation and smoke density values for electrical and optical fiber cables used in spaces transporting environmental air (i.e. plenum). |

# Appendix II: Contact addresses for Standards Organisations and other interested bodies

AFNOR

Association Française de Normalisation
Tour Europe
92049 Paris la Défense Cedex
France
Tel    +33 1 42 91 55 55
Fax   +33 1 42 91 56 56
www.afnor.fr

ANSI

American National Standards Institute
11 West 42nd Street
13th Floor
New York
NY 10036
USA
Tel    +1 212 642 4900
Fax   +1 212 302 1286
www.ansi.org

BiCSi

BiCSi
8610 Hidden River Parkway
Tampa
Fl 336337-1000

USA
Tel    +1 813 979 1991
Fax    +1 813 971 4311
www.bicsi.org

BSI    British Standards Institution
389 Chiswick High Road
London W4 4AL
United Kingdom
Tel    +44 181 996 9000
Fax    +44 181 996 7460
www.bsi.org.uk

CEN    European Committee for Standardisation
Rue de Stassart 36
B-1050 Brussels
Belgium
Tel    +32 2 550 0811
Fax    +32 2 550 0819
www.cenorm.be

CENELEC    CENELEC
Rue de Stassart, 35
B-1050 Brussels
Belgium
Tel    +32 2 519 6871
Fax    +32 2 519 6919
www.cenelec.org

CSA    Canadian Standards Association
178 Rexdale Boulevard
Etobicoke
ON M9W 1R3
Canada
Tel    +1 416 747 4044
Fax    +1 416 747 2475
www.cssinfo.com

| | |
|---|---|
| Dansk Standard | Dansk Standard (DS) Electrotechnical Sector |
| | Kollegievej 6 |
| | DK-2920 Charlottenlund |
| | Denmark |
| | Tel    +45 39 96 61 01 |
| | Fax   +45 39 96 61 02 |
| | www.ds.dk |
| | |
| DIN | Deutsches Institut für Normung e. V. |
| | Burggrafenstrasse 6 |
| | 10787 Berlin |
| | Germany |
| | Tel    +49 30 2601-0 |
| | Fax   +49 30 2601-1231 |
| | www.din.de |
| | |
| EIA | Electronic Industries Alliance |
| | 2500 Wilson Boulevard |
| | Arlington |
| | VA 22202-3834 |
| | USA |
| | Tel    +1 703 907 7500 |
| | Fax   +1 703 907 7501 |
| | www.eia.org |
| | |
| ETCI | Electro-Technical Council of Ireland |
| | Unit 43 |
| | Parkwest Business Park |
| | Dublin12 |
| | Ireland |
| | Tel    +353 1 623 99 01 |
| | Fax   +353 1 623 99 03 |
| | www.etci.ie |
| | |
| ETSI | European Telecommunications Standards Institute |
| | Route de Lucioles |

F-06921 Sophia Antipolis Cedex
France
Tel    +34 4 9294 4200
Fax    +34 4 9365 4716
www.etsi.fr

EUROPACABLE    The European Confederation of Associations of
Manufacturers of Insulated Wire and Cable
c/o CABLEBEL asbl
Diamant Building 5th Floor
Bld August Reyers 80
B-1030 Brussels
Belgium
Tel    +32 2 702 62 25
Fax    +32 2 702 62 27
www.europacable.com

FCC    Federal Communications Commission
1919 M Street NW
Room 702
Washington
DC 20554
USA
Tel    +1 202 418 0200
Fax    +1 202 418 0232
www.fcc.gov

FIA    Fibre Optic Industry Association
Owles Hall
Owles Lane
Buntingford
Herts SG9 9PL
United Kingdom
Tel    +44 1763 273039
Fax    +44 1763 273255
www.fibreoptic.org.uk

ICEA                     Insulated Cable Engineers Association
                         PO Box 1568
                         Carrollton
                         Georgia 30117
                         USA
                         www.icea.net

IEC                      International Electrotechnical Commission
                         Rue de Varembe, 3
                         PO Box 131
                         CH-1211 Geneva 20
                         Switzerland
                         Tel    +41 22 919 02 11
                         Fax    +41 22 919 03 00
                         www.iec.ch

IEE                      Institute of Electrical Engineers
                         Savoy Place
                         London
                         WC2R 0BL
                         UK
                         Tele   +44 (0)20 7240 1871
                         Fax    +44 (0)20 7240 7735
                         www.iee.org

IEEE                     Institute of Electrical and Electronic Engineers
                         445 Hoes Lane
                         PO Box 1331
                         Piscataway
                         NJ 08855-1331
                         USA
                         Tel    +1 732 981 0060
                         Fax    +1 732 981 9667
                         www.ieee.org

ISO                      International Organisation for Standardisation
                         Rue de Varembe, 1

CH-1211 Geneva 20
Switzerland
Tel    +41 22 749 01 11
Fax    +41 22 733 34 30
www.iso.ch

ITU

International Telecommunications Union
Place des Nations
CH-1211 Geneva 20
Switzerland
Tel    +41 22 730 51 51
Fax    +41 22 733 72 56
www.itu.int

NEC

Nederlands Elektrotechnisch Comite
Kalfjeslaan 2
Postbus 5059
NL – 2600 GB Delft
The Netherlands
Tel    +31 15 269 03 90
Fax    +31 15 269 01 90
www.nni.nl

NEK

Norsk Elektroteknisk Komite
Harbitzalleen 2A
Postboks 280 Skoyen
N-0212 Oslo
Norway
Tel    +47 22 52 69 50
Fax    +47 22 52 69 61
www.nek.no

NEMA

National Electrical Manufacturers Association
1300 North 17th Street, Suite 1847
Rosslyn
VA 22209
USA

Tel   +1 703 841 3200
Fax   +1 703 841 3300
www.nema.org

NFPA                 National Fire Protection Agency
                     1 Batterymarch Park
                     PO Box 9101
                     Quincy
                     MA 02269-9101
                     USA
                     Tel   +1 617 770 3000
                     Fax   +1 617 770 0700
                     www.nfpa.org

NNI                  Nederlands Normalisitie-instituut
                     Kalfjeslaan 2
                     Postbus 5059
                     NL – 2600 GB Delft
                     The Netherlands
                     Tel   +31 15 269 03 90
                     Fax   +31 15 269 01 90
                     www.nni.nl

NSF                  Norges Standardiseringsforbund
                     Drammensveien 145
                     Postboks 353 Skoyen
                     N-0213 Oslo
                     Norway
                     Tel   +47 22 04 92 00
                     Fax   +47 22 04 92 11
                     www.standard.no

SEK                  Svenska Elektriska Kommissionen
                     Kistagangen 19
                     Box 1284
                     S-164 28 Kista Stockholm
                     Sweden
                     Tel   +46 84 44 14 00

Fax   +46 84 44 14 30
www.sekom.se

SESKO          Finnish Electrotechnical Standards Association
               Sarkiniementie 3
               PO Box 134
               SF-00211 Helsinki
               Finland
               Tel   +358 9 696 391
               Fax   +358 9 677 059
               www.sesko.fi

SFS            Finnish Standards Association SFS
               Maistraatinportti 2
               FIN-00240 Helsinki
               Finland
               Tel   +358 9 149 9331
               Fax   +358 9 146 4925
               www. sfs.fi

SIRIM          SIRIM Berhad
               1 Persiaran Dato'Menteri
               PO Box 7035
               Section 2
               40911 Shah Alam
               Malaysia
               Tel   +60 3 559 2601
               Fax   +60 3 550 8095
               www.sirim.my

Standards      Standards Australia
Australia      PO Box 1055
               Strathfield
               NSW 2135
               Australia
               Tel   +61 2 9746 4700
               Fax   +61 2 9746 8540
               www.standards.com.au

SNZ          Standards New Zealand
             155 The Terrace
             Private Bag 2439
             Wellington
             New Zealand
             Tel   +64 4 498 5990
             Fax   +64 4 498 5994
             www. standards.co.nz

TIA          Telecommunications Industry Association
             2500 Wilson Boulevard, Suite 315
             Arlington
             VA 22201-3836
             Tel   +1 703 907 7700
             Fax   +1 703 907 7727
             www.tiaonline.org

UL           Underwriters Laboratories Inc.
             333 Pfingsten Road
             Northbrook
             IL 60062
             USA
             Tel   +1 847 272 8800
             Fax   +1 847 272 8129
             www.ul.com

UTE          Union Technique de l'Electricité
             33, Av. General Leclerc – BP 23
             F-92262 Fontenay-aux-Roses Cedex
             France
             Tel   +33 1 40 93 62 00
             Fax   +33 1 40 93 44 08
             www.ute-fr.com

VDE          Deutsche Elektrotechnische Kommission im DIN
             und VDE
             Stresemannallee 15

D-60 596 Frankfurt am Main
Germany
Tel   +49 69 63 080
Fax   +49 69 63 12 925
www.dke.de

Obtaining copies of Standards:
The Standards-writing bodies may be contacted directly or Standards may usually be obtained through the national standards body of the country in which you reside. They are listed above under the following headings:

| | |
|---|---|
| Australia | Standards Australia |
| Canada | CSA |
| Denmark | Dansk Standard |
| Finland | SESKO, SFS |
| France | AFNOR, UTE |
| Germany | DIN, VDE |
| Holland | NEC, NNI |
| Ireland | ETCI |
| Malaysia | SIRIM |
| New Zealand | NZS |
| Norway | NEK, NSF |
| Sweden | SEK |
| United Kingdom | BSI |

Alternatively, most standards can be purchased through:

Global Engineering Documents
Customer Support
15 Inverness Way
Englewood
CO 80112
USA
Tel +1 800 624 3974
Fax +1 303 792 2192
www.global.ihs.com

# Index